Conference on

DIGITAL INSTRUMENTATION

12 - 14 NOVEMBER 1973

Organised by the

Electronics Division of the
Institution of Electrical Engineers

in association with the

Institute of Measurement and Control
Institution of Electronic and Radio Engineers

Venue

Institution of Electrical Engineers
Savoy Place
London WC2

Organising Committee

H. A. Dorey (*Chairman*)

Dr. D. I. Crecraft

P. Cooke

B. Freeman

Prof. R. L. Grimsdale

Dr. M. T. G. Hughes

Prof. J. M. Nightingale

D. R. Ollington*

A. R. Owens

K. R. Thrower

*representing the Institution of Electronic and Radio Engineers

The cover was designed by Mike Sharp and was inspired by the waveforms of a high speed analogue to digital converter operating on the successive approximation principle.
Published by The Institution of Electrical Engineers
Printed by The Colmore Press Ltd., 62A Queens Road, Peckham, London SE15
 ISBN: 0 85296116 2

Contents

The Institution of Electrical Engineers is not, as a body, responsible for the opinions expressed by individual authors or speakers

Contents

Contents

Contents

List of Authors

DIGITAL INSTRUMENTATION

H. A. Dorey.

"The gods are just, and of our pleasant vices
Make instruments to plague us"

(King Lear)

This cry from his heart will be echoed by all those involved in measurement who have known unreliable instrumentation. Yet how many of us have brought this curse upon ourselves? Where a process is on the forefront of technology the instrumentation needed for checking it may be on the limit of feasibility.

The art of the instrumentation engineer is to bring a difficult measurement into the practical domain. The need for high frequency measurements, as is found in communications, created today's sampling techniques. With digital computers the complex information paths have brought into use diagnostic methods, where the behaviour of the computer in response to a certain stimulus is used to analyse faults. Just as with a doctor examining a patient care is needed in the interpretation of results for the answers may be ambiguous.

The inherent inaccuracy of analogue displays and the need for more accurate measurements forced digital instruments into the laboratory, before they could be made reliably.

HISTORY

The first digital electronic instruments used valves, with meter or gas discharge displays and frequently a number of mechanical parts.

To appreciate the rate of change in electronics, consider the circuit elements available for £5.

1952	A sub-miniature pentode - 1¼″ high
1956	A silicon N.P.N. transistor, f_t 15MHz
1958	A germanium power transistor
1962	A germanium 500MHz transistor
1964	A planar bistable
1968	A decade counter
1973	1000 bit store
1975	A microprocessor

At each of these dates the component would be designed into circuit with a corresponding amount of thought. It is surprising that the first digital voltmeters and counters could be built in 1952. Electronic parts were obviously very expensive, so only the amplifiers were made of valves. Voltages were switched and displays driven by stepping switches - to achieve reliability and lower the cost! An early handbook advises;- "The oil-bath feature eliminates the need for periodic lubrication of switches." I can still remember the pool of oil on my bench..........the dipstick to check the oil-level..........it was like driving a vintage motor car!

It was not only active devices that had their problems. Precision resistors were used in the form of a Kelvin-Varley bridge, in which a constant current was maintained in all resistors. If the current was switched some resistors took 15 seconds to regain their specified value, and many never reached it.

Not until 1959-60 were digital instruments made entirely from transistors, the instruments progressed from the state where you were surprised that they worked, to a state where they were accepted for production. The display still contained 50-100 filament bulbs, was the major cause of unreliability and required expensive power transistors.

In 1964 the arrival of lower cost gas discharge displays, followed later by integrated circuits to drive them, provided a major step forward. Counters and voltmeters appeared in every laboratory, and sometimes in factories.

H. A. Dorey is with the Solartron Electronic Group Ltd.

With the advent of the planar integrated circuit, data processing became economic. It was natural to incorporate these circuits in instrumentation, and they were used to calculate mean and rms voltages, or to arrange the format of printed statements. In 1968 they were used in the first correlator, and now a large proportion of instruments have a computational facility.

Processors have been designed into many instruments and some have several processors for reliability or speed. It should follow that the user will receive information that he wants in a form he can understand.............

A major computer manufacturer forcast in the mid-sixties that all instrumentation would be replaced by computers and peripheral devices. Certainly many computers have been swallowed by the instrumentation.

TECHNOLOGY

Most digital circuitry is implemented using silicon planar technology. As process control is improved and yields increase, more complex circuits become possible. The development involves considerable investment in device technology, particularly in the chemical and photographic equipment.

At the present time planar processes are mainly suitable for digital circuits since the analogue equivalent requires tighter control of each production process, and the yields are lower. It is as practical to place 10 000 digital elements on a chip as 500 analogue linear elements. For this reason designs are made to have the highest possible digital content. Development in thick and thin film circuits and improvement of analogue planar circuits may reduce the emphasis on digital circuits.

The instrument engineer does not use sufficient devices to justify his own processes. He has to accept those available.

No replacement can be seen for the planar process in the next ten to fifteen years. The most likely challenge comes from the optical field, itself helped by photographic developments for the planar process. This work is in its infancy.

Only with Surface Acoustic Waves can we find a technology that could be comparable to the planar process in its importance, yet the technologies are complementary rather than supplementary. Surface Acoustic Waves are continuous analogue signals, and not easily used for logic. Their unique property is that they may be analysed in the time domain with high accuracy. They provide an analogue process of accuracy comparable to a digital system. The production process is photographic, and could be as suitable for mass production on quartz or lithium niobate as the silicon planar circuit.

Display technologies have been covered by so many conferences that they have been omitted deliberatly from this paper.

CURRENT TRENDS

Digital circuits are replacing their analogue predecessors in the most unexpected ways. Consider an analogue phase meter, which can now be bought with analogue inputs and moving coil meter displays, but the circuitry inside is entirely digital. Because multiplication is involved, the cheapest way to achieve the required specification uses digital arithmetic facilities. The result may have higher performance or lower cost than conventional methods. It will not be long before digital displays are cheaper than the analogue meters.

Digital instruments can be made very economically if a high specification is not required. In this situation cost becomes the most important design parameter, and mass production is essential. Car instruments and household meters are suggested as possible applications, if the price can be made low enough. A digital kWh meter costing £5 is under study in several locations. This price seems very low, but the meter is simple compared with a calculator. Hand held calculators sell for under £30 now, and the price is expected to decrease to £10 in the next two to three years. The price is falling so rapidly that manufacturers give as little as 5 days notice of price-cuts to retailers.

This low cost production is not regarded favourably by all engineers, but it is introducing new production methods and new circuits that can be used to great advantages in expensive instruments. Resistive networks on ceramic substrates are very cheap, and have tightly defined inductance and capacitance. For computer designed rf circuits they are modelled easily and are the obvious choice.

High accuracy in phase and voltage measurement is achieved using time-division methods. Few relatively expensive linear components are used before the circuitry becomes digital. The answer may then be in the wrong form, but mathematical correction can be implemented easily, using calculation methods borrowed from cheaper instruments - or even calculators.

The greatest rate of growth is in digital test gear for checking the digital equipment itself. High speed computation and pulse code modulation communication are now so fast that special test equipment is essential. It may be difficult to tell if a fault lies in the equipment, the test gear, or was caused by the attachment of the test gear, perhaps affecting the standaing wave pattern. With bit rates approaching 1 Gigabit/s this problem becomes acute, and test gear has to be designed as an integrated part of the equipment. It is important that test gear should not indicate failures when it is the test that is faulty, so this dedicated instrumentation must identify its own faults.

PHYSICAL INPUTS

The ease with which digital data can be processed is creating new sensors with digital outputs. Temperature, pressure, flow, density, strain and position can now be measured very accurately - at a much higher cost. We are awaiting the parallel development of cheap digital sensors; so far only temperature and pressure are under serious consideration.

Optical position measurement using photo-sensors can now achieve 0.05% accuracy, and multiple arrays are being developed. This enables measurement of shapes and areas, and even three dimentional measurements have become possible. The use of this equipment is in its infancy, but shows considerable promise.

Digital sampling of shapes requires numerical methods of analysis. It is interesting to note how many logic designers are implementing methods suggested by Gregory, Newton or Aitken. This work was started in the mid-seventeenth century, but is still the best means of interpolation or reduction of data.

THE PLACE OF THE ENGINEER

The job of the engineer has been changing as rapidly as the instrumentation. In 1960 he designed bistables and elementary amplifiers, which are now standard parts. Today he uses an LSI pack if it is available. If he requires a special function he may model his amplifiers using a computer, or shape his logic circuits on silicon. By the standards of 1960 he appears very academic, and this shift of emphasis is expected to continue..........

The user of the instrumentation is becoming more specialised in his own field. He expects his instruments to be rugged, reliable and unobtrusive, and seldom wishes to be an electronics expert. The equipment is becoming increasingly complex, but it must still accept apparently simple inputs and give answers clearly in the form required by the user. He does not wish to be bothered with the noise and interference that could have made the reading almost impossible.

When equipment ceases to work, it should be obvious that it has failed, and it must be easy to rectify. Faults should announce their presence............

"Even like the deaf adder that stoppeth". (Psalm 58)

In these ways the skill of the designer shows in every instrument, and success depends on his knowledge of the measurements involved, and of the best methods available. The successful instrument designer is in very short supply.

DIGITAL MEASUREMENT OF LASER DOPPLER FREQUENCY.

C.A. Hobson, M.J. Lalor, and W. Weston.

Introduction

In the past ten years the unique properties of laser radiation have begun to be utilised for the measurement of fundamental parameters in engineering. In particular the continuous wave gas laser, which provides coherent monochromatic visible radiation, has been used to make point velocity measurements in fluids. If a source of acoustic or electromagnetic waves of constant frequency, f, is put in motion, the frequency of the waves received by a fixed observer will change. This is the well known Doppler effect.

A similar phenomenon is observed when the source is fixed but the waves are scattered or reflected from a body in motion before being received by the fixed observer. In either case the apparent change in frequency - the Doppler shift, fd - is proportional to the velocity. Conventional light sources have a bandwidth much greater than the Doppler shift at reasonable velocities. Thus the change in frequency, relative to the source frequency, is usually masked by the spread in the source frequency except in rare cases of very high velocities. However a single axial mode of a He-Ne laser operating at 4.71×10^{14}Hz has a bandwidth of about 10 Hz, so although the Doppler shift of light from a laser source caused by motion of an emitter or reflector moving at moderate speeds is still small compared to the source frequency, it is large compared to the source bandwidth.

The Doppler shifted light still has a frequency ($f \pm fd$) in the order of 10^{15}Hz. This is too high for electronic measurement and if it is desired to measure this frequency directly an optical technique such as a scanning plate Fabry-Perot interferometer (1) must be used. The Doppler shift can be reduced to the range of electronic measurement by an optical hetrodyne technique in which the Doppler shifted scattered light is mixed with an unshifted reference beam, from the same laser, on the photocathode of a photomultiplier tube. This results in a photocurrent that contains a beat frequency at the frequency of the Doppler shift. In the experiments described below this is in the range of 20 - 200 kHz.

A typical experimental arrangement is shown in fig. 1a. (2). A simplified arrangement was suggested (3) in which two equal intensity laser beams were brought together by a single lens to a focus inside the fluid under study. Fig. 1b. The frequency of light scattered from the probe volume may be interpreted either as a differential Doppler signal (4) or in terms of the rate at which a scatter particle crosses alternate bright and dark interference fringes. In either system the frequency fd is related to the velocity, v, of a scattering particle passing through the measurement volume by $v = \lambda f/2 \sin(\theta/2)$, where λ is the wavelength of the laser light and θ is the angle between the two laser beams.

C.A. Hobson is with the Department of Electrical and Control Engineering, Liverpool Polytechnic.
M.J. Lalor and W. Weston are with the Department of Mechanical, Marine and Production Engineering, Liverpool Polytechnic.

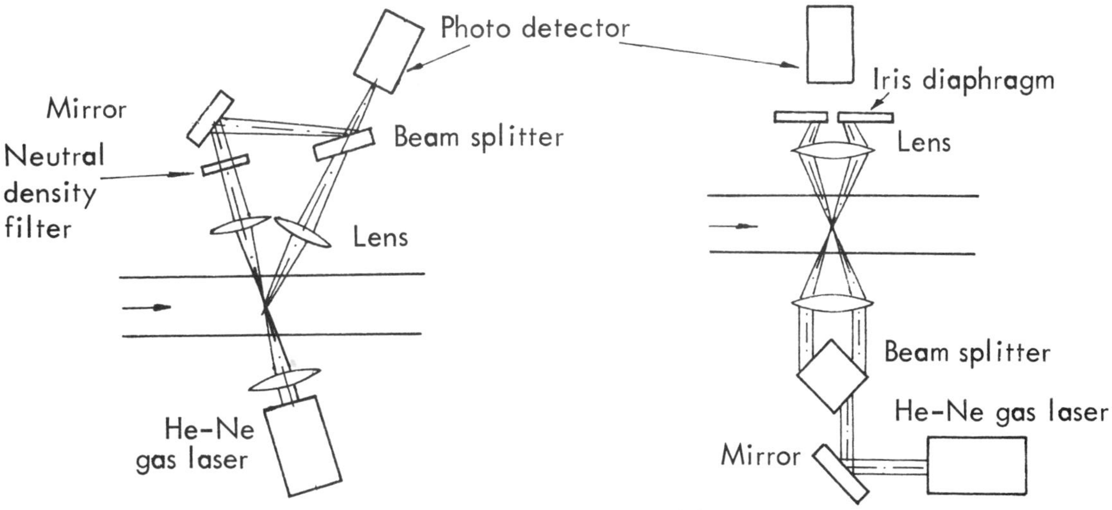

(a) Doppler Mode (b) Fringe Mode

Figure 1: TYPICAL OPTICAL ARRANGEMENTS.

The Doppler signal is only present when a scattering centre crosses the measurement volume, therefore the output from the photomultiplier tube is not a continuous signal. Drop-out, which is the proportion of time that a signal is not received, can vary from as much as 98% in some fluid systems to almost zero in systems for measuring the velocity of a solid surface. Fig.2 illustrates typical waveforms after the photomultiplier signal has been band-pass filtered.

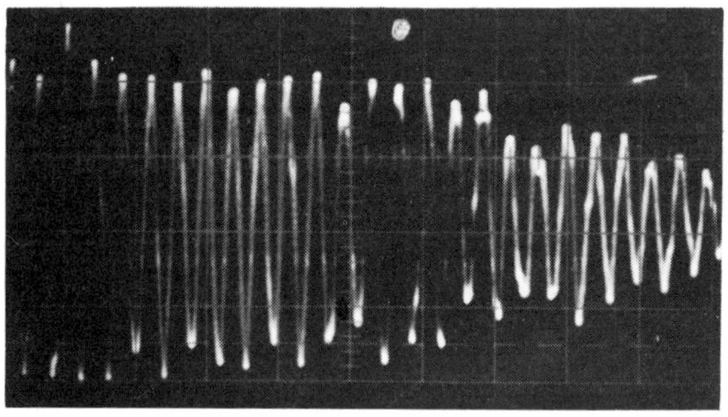

IV/cm @ 20 μs/cm.
(a) Low dropout (Solid surface)

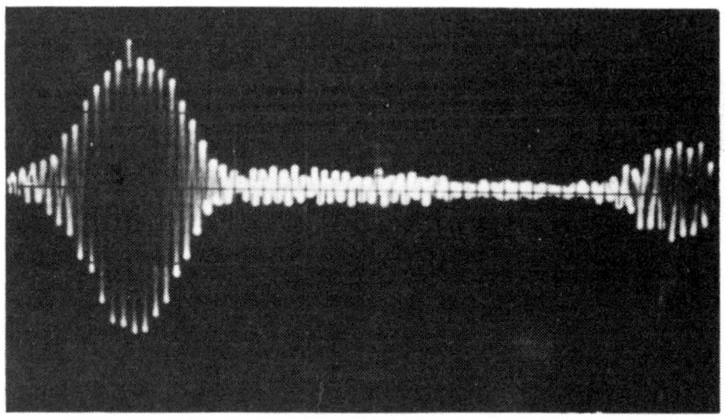

IV/cm @ 200 μs/cm.
(b) High dropout (Fluid)

Figure 2: TYPICAL FILTERED WAVEFORMS FROM PHOTO MULTIPLIER.

Various techniques for the measurement of the Doppler frequency have been described. These include the use of a spectrum analyser (5), an autodyne tracking filter with a freeze system (6) and the gating of a conventional frequency meter by the Doppler signal (7). In high drop out systems operating with either a closed loop or varying flow rates these techniques are not satisfactory due to the difficulty of obtaining suitable outputs, or to the time required to make measurements.

A Digital Doppler Frequency Meter

Based on experience with a prototype (8) and on the need for a versatile piece of equipment, it was decided to design a meter capable of accepting signals derived from high or low drop-out systems, and capable of operating independantly of the number of fringes that a scatter particle passes through. The output was required to be an analog signal for use in a feedback system. To cater for future computer processing requirements it was decided that digital outputs should also be available. A system, whose block diagram is shown in fig.3 was arrived at.

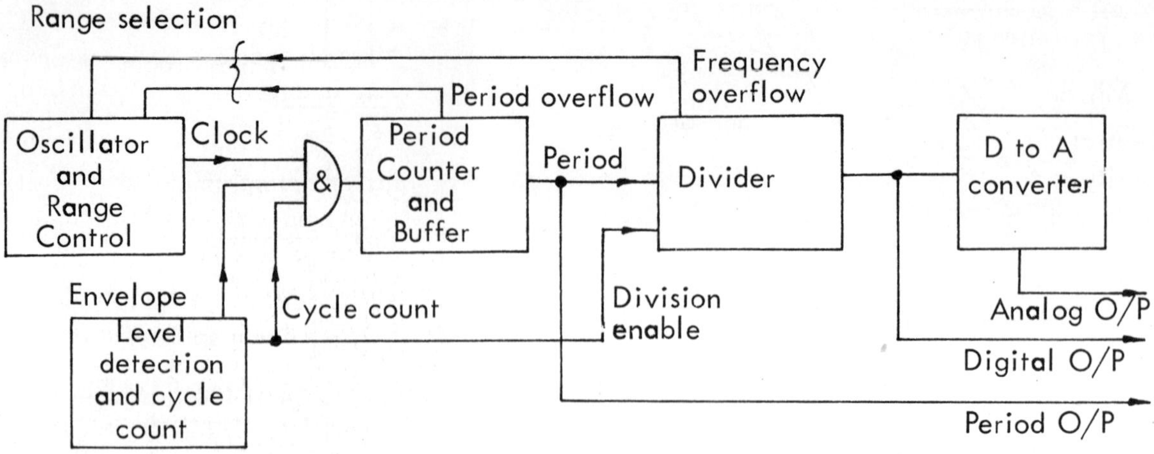

Figure 3: BLOCK DIAGRAM OF DOPPLER FREQUENCY METER.

Frequency measurement is carried out by the indirect method of measuring the time taken for the Doppler signal to complete an integral number of cycles, and then calculating the frequency. Since the Doppler signal is not of constant amplitude it is necessary for measurements to be carried out between zero-crossing points on the waveform.

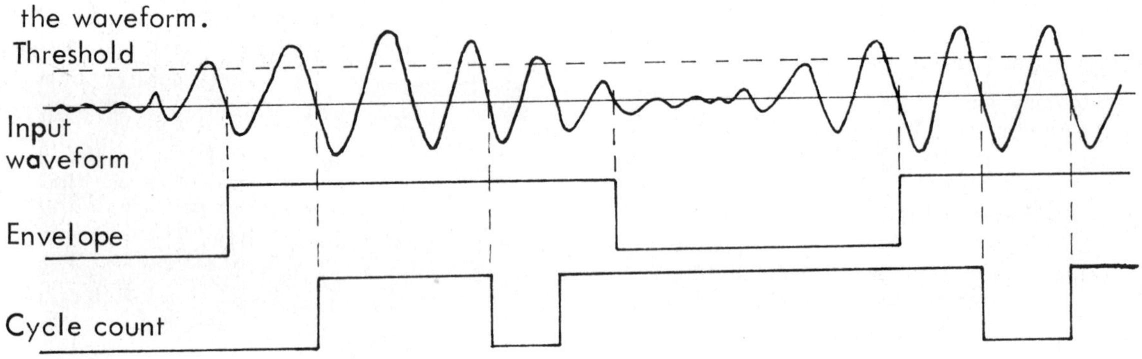

Figure 4: CONTROL WAVE FORMS.

The incoming Doppler signal is applied to a circuit which generates a rectangular envelope around the signal as shown in fig.4. The envelope is present from the first zero-crossing of the Doppler signal after its amplitude exceeds a threshold voltage, until the first zero-crossing after its amplitude falls below the threshold voltage.

When the envelope signal is present, a binary counter (the cycle counter) is caused to increment by 1 on each positive to negative transition of the Doppler signal. The output of the cycle counter is gated so that a pulse, whose duration is equal to N cycles of the Doppler signal, is generated. N is selected by a switch to be either 1, 2, 4 or 8. In fig.4, N is assumed to be 2.

A combination of the envelope and cycle count pulses are used to gate a crystal oscillator output into the period counter. At the end of the cycle counter pulse the contents of the period counter are proportional to the average period of the Doppler signal, regardless of the number of Doppler bursts required to complete the period counter sequence.

When a measurement is completed, the contents of the period counter are loaded into a buffer and then applied to a binary divider. Division is completed in 12 cycles of a synchronising clock signal, and the resulting digital output is applied to a D to A converter.

Range selection is either manual, from a front panel switch, or automatic by detection of two successive overflows from the period counter or divider. The need to detect two successive overflows reduces the risk of false operation due to the input circuits responding to noise. When on manual range selection the overflows are used to indicate when a change of range is required. In operation this has been found to be a most useful feature since it assists in the setting of the threshold voltage above noise but below the Doppler signal amplitude.

Level Detection and Cycle Counter

Since the operation of the entire instrument is based on the level detection and cycle counter, these circuits will be described in some detail. Other circuits in the system are of fairly standard design and will not be described.

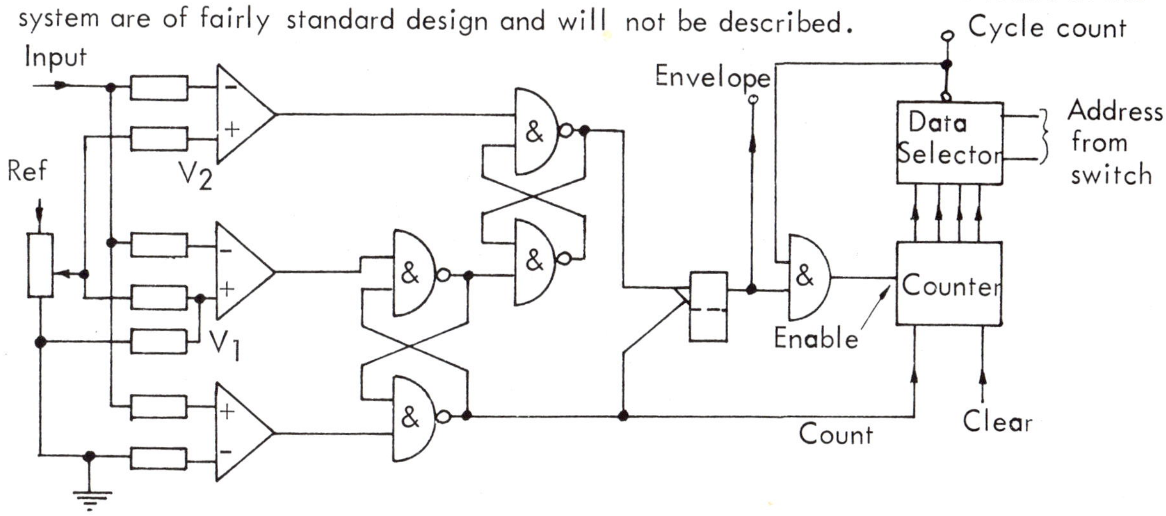

Figure 5: LEVEL DETECTION AND CYCLE COUNTER.

Fig.5 shows the essential features of the circuit. The Doppler signal is applied to three high speed comparators each of which has a different reference voltage. The comparators drive two latches. For signals of instantaneous amplitude greater than V_1 the lower latch is set and for signals of instantaneous amplitude greater than V_2 both latches set. The latches are reset as the input signal goes negative. If the top latch has been set the D-type flip-flop will be clocked to the 1-state, but if only the bottom latch has been set the D-type flip-flop will be clocked to the 0-state.

The ratio, V_2/V_1, is a compromise. V_2 must be above the amplitude of the noise on the Doppler signal, or false triggering of the D-type flip-flop will occur. If V_1 is too close to V_2 it is possible to have false operation at the trailing end of the Doppler burst since the resetting of the D-type flip-flop requires that there should be at least one cycle of amplitude greater than V_1 and less than V_2.

One output of the counter is selected and inverted by the data selector. If either the D-type flip-flop output (the envelope) or the data selector output (cycle count) goes to 0, then the counter will cease to increment. Clearing of the counter is synchronous and therefore the count returns to zero on a positive to negative transition of the Doppler signal.

Instrument Accuracy and Limitations

The main sources of error in the measurement system derive from the quantisation of period and frequency data. Denoting P as the period counter contents the design required $80 < P < 1024$. If P falls outside this range there will be an overflow from either the counter or divider.

The error due to quantisation of P has a maximum value of 1.25% at lower values of P. This error could be reduced by increasing the lowest possible value of P, but to do this would require the system to operate either over more cycles of the Doppler signal, or with a higher frequency oscillator. With high drop out signals measurement over more than 8 cycles is not practical, and an 8 cycle measurement with the limitations specified above requires an oscillator frequency of ten times the Doppler frequency.

The frequency is represented by the contents of an 11-bit quotient register, which stores the value of frequency, $F = 80/P$. The quantisation error in F is a maximum when P is a maximum. The error in F could be reduced by use of a longer quotient register, but a D to A converter with more inputs would be required. This was not considered to be justified due to the accuracy limitations of the circuits driven by the analog output.

Filtering does not remove all noise from the photomultiplier signal. The input circuits have hysteresis so that they do not respond to the noise but nevertheless the noise causes an apparent change of period. By measuring the time required for the Doppler signal to complete a number of cycles the effects of noise are reduced.

The time taken for a measurement to be completed increases as the period of the Doppler signal increases. In some applications it may be necessary to allow for this difference in measurement time.

Applications

Two applications are outlined below. They conform to:

1. The high drop out mode in which the Doppler signal is used to control the flow output of an hydrostatic system,

and 2. the low drop out mode where the processor is used to measure slip in the rolling elements of a bearing.

In the former the laser Doppler velocimeter is used to accurately control the flow output from a two stage electrohydraulic servo valve (8). This can then, by closing the local flow control loop round the valve, define the position of an actuator in a hydraulic control system.

Alternative methods of flow or velocity measurement are not suitable here since they either have a poor frequency response, or they introduce a disturbing element into the flow. High integrity hydraulic systems typically involve fluid filtering of better than 5 microns. The resulting low density of scattering particles results in a need for an instrument whose characteristics are similar to those of the one described.

Measurement of the speed of a bearing cage, however, is comparatively easy, but the measurement of the nominally epicyclic speed of the rollers themselves is more difficult. This measurement is of interest because slip of the bearings is thought to be related to wear rate and is also a source of power loss in situations where a bearing is lightly loaded and running at high speed such as in a gas turbine. Various techniques have been tried (9, 10) involving, for example, the drilling of small holes in the rollers and the use of an inductance pick-up. Preliminary work on the measurement of roller bearing slip using a laser Doppler method has been reported by one of the authors (11). Additional experiments using the digital frequency measurement instrumentation described are continuing and it has been found to be ideal for this application.

Conclusions

An instrument capable of operating with both high and low drop-out signals has been developed. The applications for which the instrument has proved satisfactory include the measurement of slip in roller bearings and the feedback of velocity information in a fluid velocity control system.

Acknowledgements

We wish to thank Mr. L. Hale and Mr. A. Stirrup for their constructional work.

References

1. Jackson, D.A., Paul, D.M: Measurement of supersonic velocity and turbulence by laser anemometry. J. Phys.E. Scientific Instruments 1971 4 173-177.
2. Foreman, J.W. Georges, E.W. and Lewis, R.D: Measurement of localised flow velocities in gases with a laser Doppler flow-meter. 1965 7 (4) 77-78.
3. Rudd, M.J: A new theoretical model for the laser Doppler meter. J.Sci.Inst. 1969 2 55-58.
4. Durst, F and Whitelaw, J.H: Optimization of optical anemometers. Proc.Roy.Soc.A 324 157-181 (1971).
5. Weston, W, Lalor, M.J: Laser velocimeter boosts reliability of hydraulic servos. Design Engineering Feb.1972. p.50-51.
6. Durrani, T.S. Greated, C, Wilmhurst, T.H: An analysis of simulators and tracking systems for laser Doppler velocimeters. Conference on Electro-optic systems in flow measurement. University of Southampton. Sept.1972.
7. Non-contact velocity measurement. Electronics today. International Aug.1972 p.55-57.
8. Weston, W. Lalor, M.J. Cleaver, J.W: Pulsatile flow measurement in hydraulic systems using a laser velocimeter. Opto-electronics 4. 1972. p.412-427.
9. Boness, R.J: Cage and roller slip in high speed roller bearings. J.Mech.Eng.Sci. 11 2 1969.
10. Higginson, G.R: Measurement of roller speed in roller bearings. I.Mech.Eng. March 1968. Reports 1-28 Experimental methods in tribology.
11. Bremble, G.R. Cleaver, J.W. and Lalor, M.J: A technique for measuring the speed of roller elements in a roller bearing using a laser velocimeter. Tribology Aug.1971 147-149.

DIGITAL SAMPLING METHODS APPLIED TO SHAFT VIBRATION ANALYSIS

P.M. Quilter and P. Cooke

Introduction:

This paper looks at some new techniques that enable the vibrations of a rotating shaft to be more accurately monitored; they result in a system interface that is naturally digital. Some or all of the techniques are also applicable to other systems which involve repeated cycles, or sinusoidally excited transducers.

Fig.1 shows the basic system schematic. Inductive displacement transducers are followed by special analogue circuitry that is digitally strobed. The sampling time is determined by a coincidence pulse which is obtained by comparing the instantaneous shaft angle with a digitally pre-set angle. For just one cycle of the carrier frequency the signal is integrated and a pre-determined number of samples can be accumulated, if an average value is required. This is useful for minimising the effects of random electrical or mechanical noise.

Because the adopted system does not use classic filters their phase-shift, which appears as an ill defined error of perhaps 60° in shaft position, has been removed. Additionally the analogue circuitry improves temperature performance, frequency response and accuracy.

A digital read-out of the displacement at a digitally pre-set angle is the users vision of the system. By manual or automatic manipulation of the digital inputs most of the test data required by a mechanical researcher can be displayed as testing proceeds. This compares to current methods where photographs must be taken and measured. Additional features include such things as the ability to measure displacement on alternate revolutions which, in a four-stroke engine, are different.

System considerations

Pairs of small inductive transducers embedded in the bearing are the most appropriate form of position sensor. The transducers are in thermal contact with each other, but only one is liable to inductance change with shaft movement. The dummy transducer is used in an attempt to reduce the effects of temperature, for the resistances change by about 30% between room and operating temperature. This compares with a need to ignore resistive effects whilst resolving inductive changes of 1 in 10^4.

A system typical of current methods is shown in Fig.2. A straightforward bridge arrangement is driven at a suitable frequency, and the output signal is processed by a band-pass filter, a phase-sensitive detector, and a low-pass filter. One channel of a C.R.O. is fed by this output signal and the other is fed by a T.D.C. marker.

Such systems have a number of disadvantages, viz:-

P.M. Quilter and P. Cooke are in the School of Applied Sciences at the University of Sussex.

a) Chart recorders are too slow, so C.R.O. traces must be photographed.

b) The C.R.O. sweep must be adjusted to cope with speed changes.

c) The T.D.C. marker is not in registration with the signal due to phase-shift introduced by the filters.

d) Temperature effects cause large drifts in the zero output signal.

e) Readings are inaccurate and not computer compatible.

By current driving the transducers, correlating the difference signal and by digitally controlling the sampling of the output signal of the integrator, superior performance has been achieved in a system that has digital inputs and outputs.

Angular position indicator

Sampling pulses are initiated when the instantaneous shaft angle becomes equal to a digitally pre-set figure. These coincidence pulses are emitted by a parallel digital comparator (See Fig.1).

Two optical pick-offs are aimed at marker-stripes on the shaft; from their output pulses the absolute shaft position is deduced. One unit gives one pulse for every θ° of rotation, e.g. 360 stripes give one pulse per degree of rotation. These pulses are digitally counted to give an indication of the number of degrees after T.D.C. To ensure that the digital output is 000 at T.D.C., the second pick-off is arranged to produce a single re-set pulse at that moment of time.

Current driven bridge

At the typical operating temperature of engine bearings the transducers resistance is about 30% higher than at room temperature. This fact limits the accuracy and zero-drift of the measurement system.

Even with transducer pairs as near identical as possible trimmers must be used to obtain a null from the bridge configuration of Fig.2 . This trimming resistance remains at room temperature and does not track with transducer temperatures; thus, if for no other reason, the bridge output will be temperature dependent. Only by rejecting signals due to resistive unbalance can a temperature dependent zero be avoided.

Furthermore an increase in transducer resistance reduces the drive current and hence the sensitivity, or scale-factor. More importantly the phase-angle between drive voltage and current is also changed; this eliminates the possibility of using a phase-sensitive detector to separate the wanted inductive signal components from the unwanted, and much much larger, resistive signal.

The circuit of Fig.3 overcomes the basic problems discussed above. It shows that the circuit is essentially a current driven bridge.

Implementation requires very considerable attention to detail. Both of the voltage controlled current sources must give virtually identical currents over a reasonable range of ambient temperature and in the face of large variations in transducer impedance. Also the phase differences must be small and constant (less than $\frac{1}{3}$ degree in our system). Similarly attention must be given to the differential amplifier for, unlike an operational amplifier, it must give a known gain and still be truly differential.

In the current driven bridge there is a constant phase angle between the drive-current and the inductive (and resistive) component of the output signal. Thus a detector sensitive only to signals in quadrature with the drive-current will reject changes in resistance, and detect changes in inductance.

Signal detection

Functionally the detector must process the difference signal from the bridge so as to generate an output voltage level which, at any instant, is equal to the change in transducer inductance. Clearly the measurement should be made in as short a time as possible, for this factor determines the rate at which the apparatus responds to a sudden change in shaft displacement.

In the chosen system inductance is measured in a single cycle of the carrier frequency of 10 KHz. This corresponds to about $3\frac{1}{2}^{\circ}$ of angular rotation at 6,000 R.P.M.; it compares with a resolution of several tens of degrees for the previous system.

How a switched rectifier processes signals is reviewed in Fig.4. No attempt has been made to show how the output level would follow a change in displacement. Instead cross-hatching has been used to simply show how the input waveform would have equal areas above and below the steady-state output level. Notice that the quadrature signals are rejected but that odd-order harmonics lead to an output signal.

Better phase-sensitive detectors can be based on a four quadrant multiplier. They can be considered as a linear version of the switched rectifier or bang-bang system. With a sine-wave reference the multiplier output will, after integration, be in Fourier integral form. It will also be in the form of the correlation integral.

Figure 5 reviews the waveforms for various inputs and shows typical integrator outputs if the multiplier is switched into the integrator for a single whole period. At the end of the integration period the integrator is ready for sampling but it can be held by delaying the re-set.

Direct signal averaging

By letting the integrator accumulate the result of n successive samples at the same angle, an average displacement can be obtained. Fig.1 shows how this is accomplished; a loadable counter is used to regulate the number of samples taken before the integrator is read and re-set.

Conclusion

The current need to use a C.R.O. and photographs of its traces as the primary measuring tool, provided the initial stimulus to devise a digital system. Faced with the need to retain inductive sensors an early decision was to use sampling methods; this approach has been fully justified by the results achieved. The digital nature of the system is also advantageous.

Additional improvements to the system accuracy are now limited by the transducers themselves. Basically the problem is to increase the ratio between sensitivity and temperature effects. Some small improvements have already been made, and further work continues.

System specifications have been improved with respect to temperature stability, accuracy, bandwidth and both angular and positional accuracy. For instance the bandwidth has been increased from 1 KHz to 5 KHz. Also it appears (it is difficult to know!) - that a stability of 10^{-4} inches has been achieved over the temperature range 20-100°C. This is an order of magnitude better than is possible with our original conventional system.

With the digital interface there is a much greater convenience of read-out, and automatic control and data-logging is a straightforward matter. However even with a manual interface the user has better facilities. Furthermore small logical sub-systems can add such extras as the ability to look at alternative cycles.

Other systems which share common features - such as carrier excited transducers or rotating shafts - may be worth reviewing in the light of the techniques discussed; but that is outside the scope of this paper.

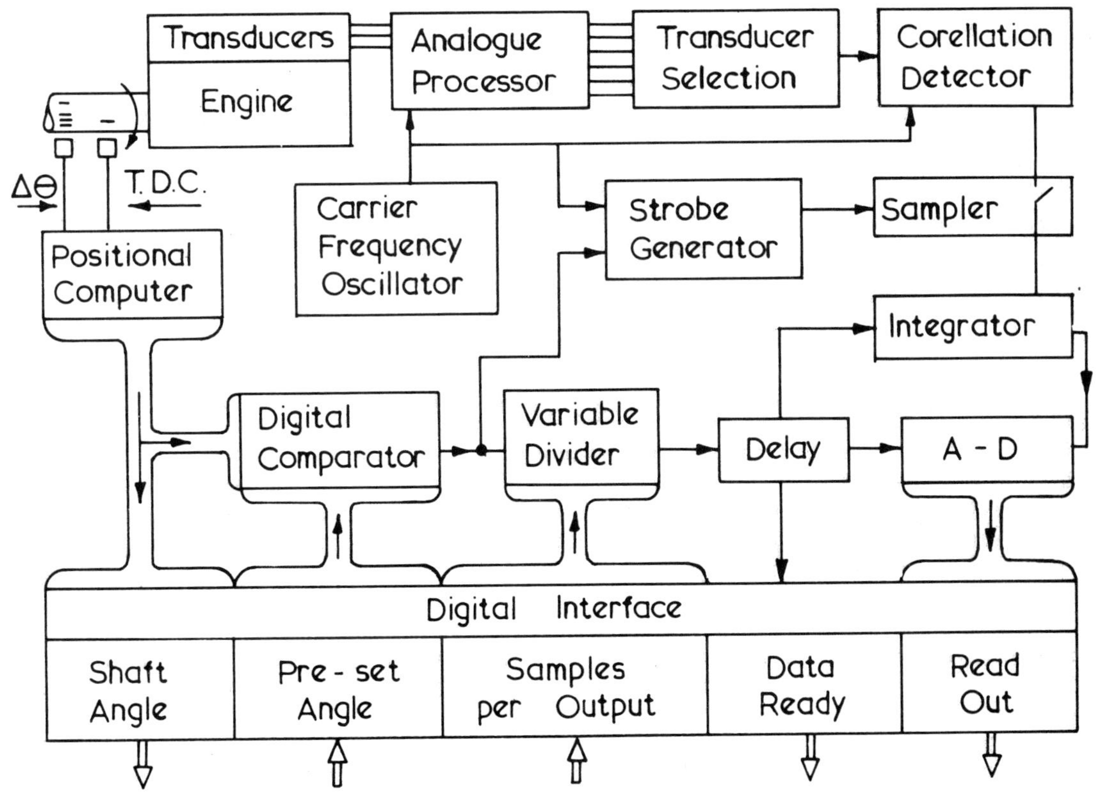

Fig. 1 Schematic of Basic System

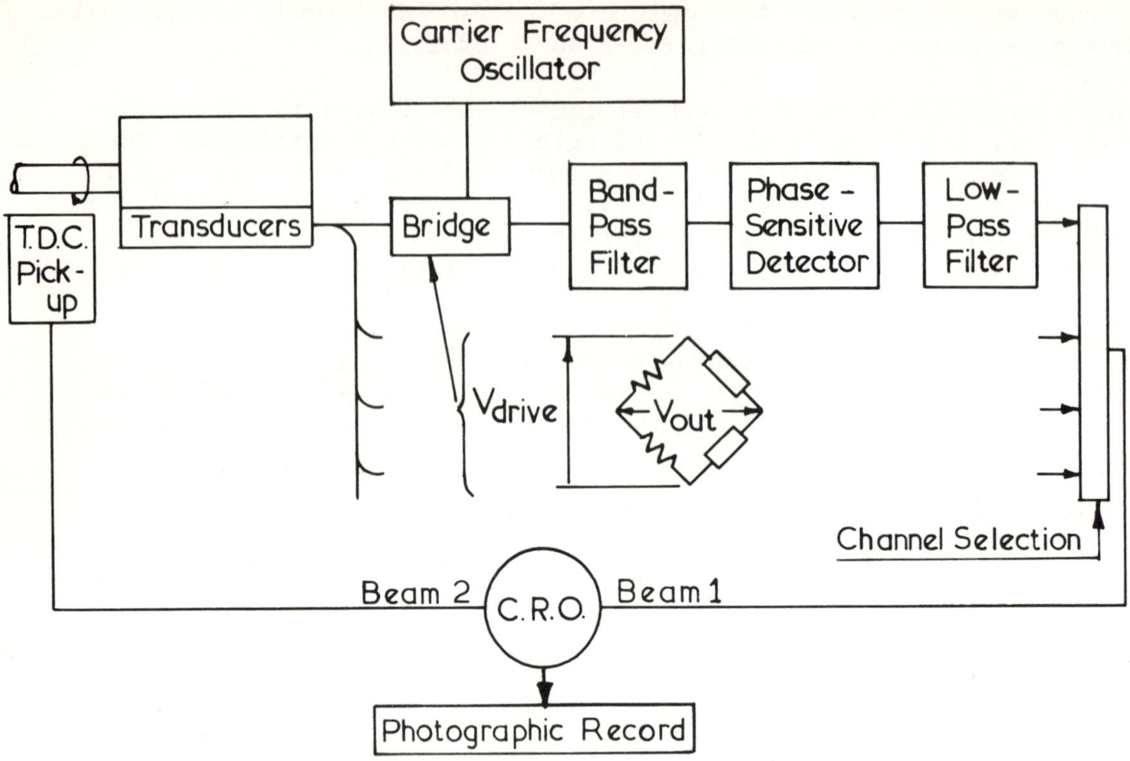

Fig. 2 Conventional System

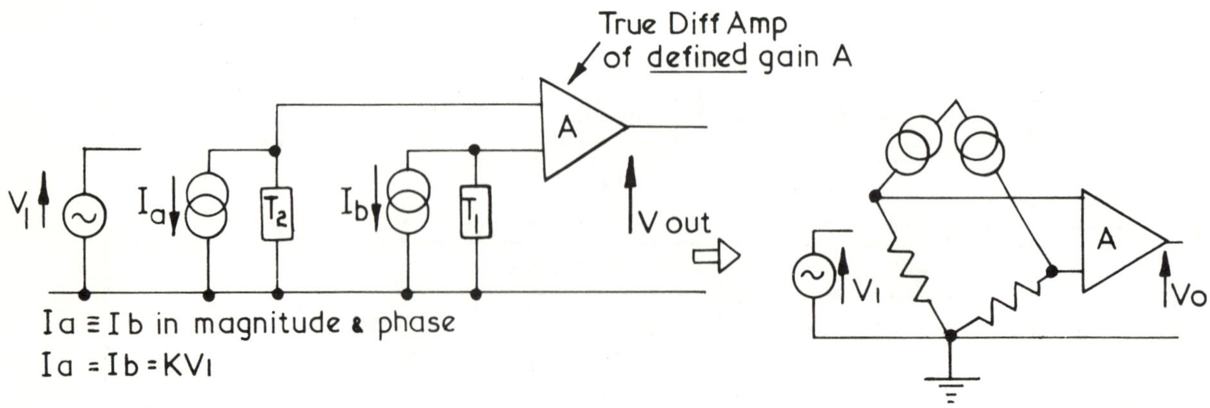

Fig. 3 Current Driven Bridge

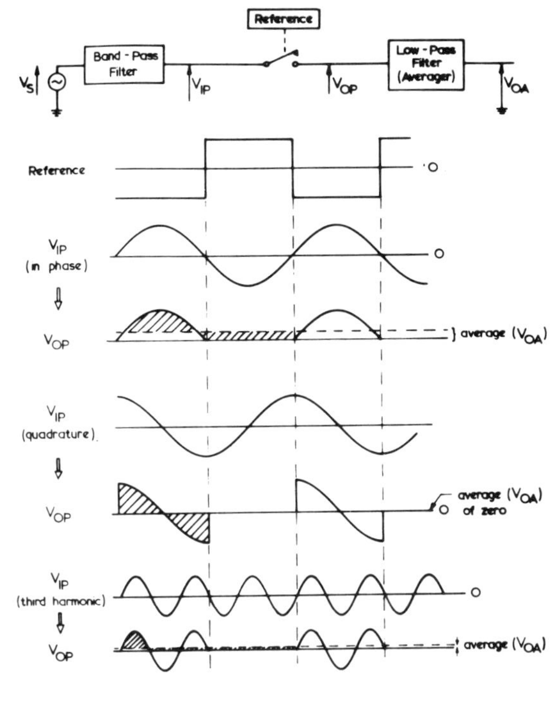

Fig. 4 Waveforms in Switched Rectifier Detector & L.P. Averager

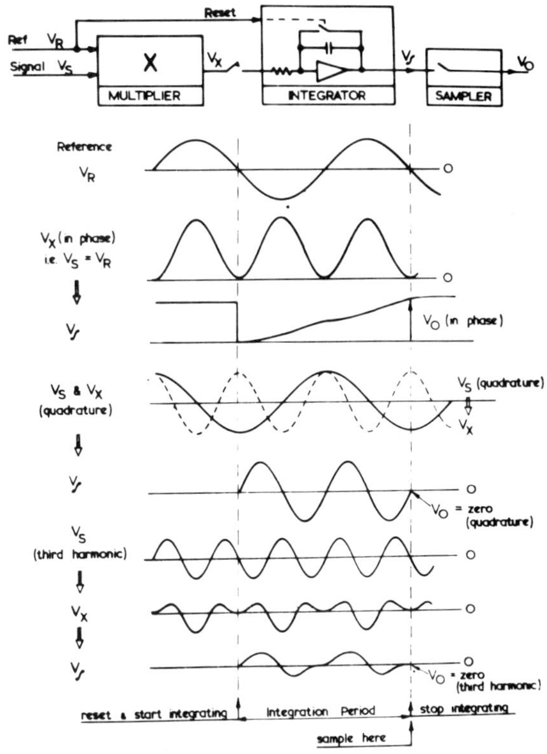

Fig. 5 Waveform for Sampled Corellator Detector

DIGITAL COMPENSATION OF SERVOMECHANISMS.

A.G. Booth, A.D.G. Hazlerigg and D.J. Woollons.

1. Introduction:

This paper describes apparatus that has been constructed over the past year to provide a facility for testing and comparing experimentally the feasibility and performance of various types of digital compensation of medium power servomechanisms.

Broadly speaking, there are two approaches to the provision of digital compensation. One is to construct special purpose digital equipment which is tailored to the particular control task. The second is to make use of a general purpose digital computer, of the type that is readily available and to adapt it to the control task through the provision of appropriate software. One of the main objectives of establishing the facility to be described is to be able to provide tested criteria by which a selection can be made between these two approaches. Factors affecting this decision include cost, technical feasibility in terms of speed of response, flexibility in setting up and tuning control laws, accuracy of control and reliability.

Speed of response (bandwidth) at reasonable cost is one of the principal advantages that special purpose equipment has over the general purpose type of digital computer. This stems from the ability to provide parallel processing of data economically for a number of restricted operations,hence avoiding the inherent slowness of a central processor through which all logical and arithmetical operations must be routed.

The present investigation has focussed attention on two types of digital machine. One is a conventional general purpose computer which has over the last four years been fitted with an extensive interface system for use online in the laboratory. The other is a special purpose, custom built, digital differential analyser type of computer capable of high rates of integration through use of incremental transfer of data between a large number of processing units operating in parallel. Both machines use a 16 bit word length, the cycle time of the former being 750 ns, the iteration period of the latter being 500 ns.

These two types of machine are being used in the laboratory to control on-line, in real time, a servomechanism of approximately 1 kW rating and having several highly resonant modes in the region 5 Hz to 15 Hz as a result of compliant couplings. Figure 1 shows a schematic of the overall system.

2. Digital Compensation of servo-mechanisms:

A servo is a device for automatically controlling the position or velocity of a mechanical system. As such, therefore, it is not naturally a discrete system. Digital compensation mainly arises in those cases where

The authors are in the School of Applied Sciences at the University of Sussex.

continuous analogue control has become inadequate in meeting stringent accuracy requirements. Accurate control requires accurate measurement, and accurate measurement is often done by counting precision states (progress through a sequence of accurately known states). The closer the sequence of states the more precise is the measurement. Once data at such a critical stage of a control loop as the formation of the error has been digitised it is natural to continue as far as possible with the same format, especially if advantages besides accuracy are potentially obtainable. For example, problems of temperature drift and component tolerances , so troublesome in analogue control loops, can be avoided almost completely by using digital techniques.

Design of linear digital compensators presents no particular theoretical difficulty, and there is a wealth of experience reported in the literature (1), using z-transform, root locus and Nyquist techniques on the one hand and difference equation methods on the other. From a practical point of view one of the main difficulties of achieving good performance with digital compensation is that of handling a sufficiently high sampling rate at reasonable cost. In the present investigation this is to be achieved on the one hand by taking advantage of incremental transfer and parallel processing, and on the other by efficiently programming a fast general purpose computer fully dedicated to the on-line control task. Although for many years the process industries have employed digital computers (GPC) for on-line control they have had the advantage of plenty of time in which to carry out the control computations as a result of relatively long process time constants. This is not the case with servo-mechanisms in which the time constants are several orders of magnitude smaller. Consequently it is by no means even certain that GPC are able to cope with the bandwidth requirements.

3. Incremental transfer signal processor (ITSP)

The special purpose digital data processor to be used as compensator in the control loop is based on the same principles as the well known digital differential analyser (2,3). Data concerning time dependent controller variables is stored in registers in the form of 16 bit words (15 bit + 1 sign bit) using twos complement format to represent negative values. Changes in the values of the variables are restricted to one increment in each processing cycle, which in the present design takes 500 nsec. i.e. the variables are updated at 2 MHz.

The ITSP is of modular construction with a minimum number of standard type modules. Programming the processor to realise a particular control law whether it be linear or non-linear, time dependent or constant, is achieved simply by physically inter-connecting the required modules. Thus simplicity is obtained at the expense of flexibility, the hallmark of the GPC, but it also goes hand in hand with high speed processing at relatively low cost. In addition, by employing DDA principles, a wide range of control functions not restricted to the linear constant coefficient class is available. This makes the ITSP a powerful research tool for investigating high performance digital servos.

The range of standard modules are designed around a basic unit consisting of two registers, the Y and R registers, and an arithmetic unit which adds the value of Y to the current contents of R on receipt of a control signal $\pm \Delta X$. The control signal may have either sign, negative being, in effect, an instruction to subtract. The Y and R registers are of the same capacity N (in the present design N = 16). As a result of each addition the contents of R may or may not overflow, i.e. exceed N. When an overflow occurs it is stored in an overflow register ΔZ. Overflow may be zero or of either sign, so a two element binary code is necessary to represent ΔZ. The contents of the Y register may be altered by up to seven ΔY inputs, each of which is in a ternary field. All ΔX, ΔY and ΔZ incremental variables are on the same ternary field and so are compatible. For example a ΔZ output from one unit can be used as the ΔX input to another unit. The basic unit performs the function $\Delta Z = Y.\Delta X$, in which ΔZ and ΔX are incremental variables, and Y is a number in the range 0 to 2^N-1. In effect it is the principal type of functional unit of a DDA type of computer.

The ITSP may contain any number of such units depending on the complexity of the control task, together with a number of specialised modifications of the basic unit for parameters and for interconnecting with the servo. For example multiplication by a parameter is effected by a basic unit from which the Y accumulator has been omitted. The parameter value is set in the Y register and ΔX and ΔZ channels operate normally. Thus a ΔZ increment is obtained after every Y/N ΔX increments. Also formation of the control error (difference between demanded value of the output position variable of the servo and its actual value) is done by a standard unit in which the ΔY accumulator is replaced by a full adder which loads Y with the difference between 'N' bit words presented at two 'N' channel ports. The number of ΔX inputs per ΔZ output is then a measure of the error. The unit also has the facility of allowing adjustment of the error significance, i.e. the gain of the unit is adjustable. For example an error of one unit in the least significant position can be made to cause one ΔZ output to occur for every ΔX input. This unit provides a convenient method of changing the static loop gain of the servo in addition to forming the error. Finally the output of the ITSP must be made compatible with the input of the power amplifier driving the servo-motor. This is achieved by means of a pulse rate to voltage amplitude convertor, which is capable of accepting a ΔZ output sequence from any of the standard processor units.

The ITSP can be operated autonomously or under the supervision of the GPC described in section 4. When operating autonomously the available controls include read and write push buttons for all Y and R registers, BCO thumb-wheel switches for setting words into registers, a BCO lamp display, a local/remote time select switch, and a run/halt switch. Data in the ITSP can be monitored readily during control experiments by on-line connection to the GPC and display system. Use of the ITSP for servo compensation is straightforward. For example, a linear compensator, figure 2, with two poles and two zeros can be designed using standard methods. The required inter-connection of functional units would then be patched and values of parameters would be set into the appropriate registers. The compensator would then be ready for use. Tuning of the compensator during tests is easily done by resetting the parameter registers. This can be done while the servo is in operation, and offers great scope for adapting a servo to changing conditions. Once satisfactory performance has been achieved the digital nature of the compensator guarantees that it can always be reset to precisely the required state on subsequent occasions, without having to make allowances for the component aging and temperature drifts normally associated with continuous analogue controllers.

4. Compensation using the general purpose computer (GPC)

In addition to the ITSP the servo-mechanism may also be controlled from a general purpose digital computer and graphic display system. These are illustrated in figure 1. The computer and graphic display are equipped with a general purpose interface suitable for connection to on-line experiments. This is employed to enable control data to be transmitted to the servo-mechanism and to allow positional information to be read from it. Such information is exchanged between the computer and registers in the input and output selectors as parallel 16 bit words. Selection of the source or destination of this data is made using address bits which are output from the computer in the same interface cycle as the data, and control of the transfer is achieved using an asynchronous hand-shake technique.

Output data can be used either as reference information to be applied to the servo input or as parameter information to enable the controller coefficients to be established or changed. In the former situation the parallel word received from the computer is converted to an ITSP compatible signal (two bits at transfer time) by using standard ITSP elements described in 3. The output from these can then be applied to the frequency to analogue convertor and motor power amplifier. In the latter case the parallel output word is used directly to establish parameters in a standard element. Thus the ITSP can accept data, which may act as either process variables or as controller coefficients, from the computer.

Input information can be read from the digital shaft encoder output or from a set of key switches and control of the iteration rate of the control algorithm can be achieved by using an external timing source to define the interface transfer handshake period. Computer outputs can be displayed on a set of lamps and these, together with the key switches, allow the operator to control the system at a point remote from the computer. The resultant facility is both powerful and flexible.

A diagram of typical programme elements is shown in figure 3. The on-line software system consists of a single programme written in assembly language and containing a group of modular routines having various individual functions and flexible inter-connection facilities. Programme parameters may be set, software altered and routines initiated by the operator using the controls and display in the vicinity of the servo-mechanism. In addition, bulk loading of parameters may be performed by paper tape input at the computer console. The two main functions are the controller and S.H.M. routines which can both be operated alone or subordinate to the plotting routine. Thus the range of experimental formats include closed loop control of the servo, transient response measurement, transfer function analysis and plotted response to sinusoidal input.

The control routine evaluates the algorithm for a linear controller in a recursive form, using first order step-wise integrations as the dynamic operations. The state memories and error signal are handled with double precision (2 x 16 bits) and comprehensive saturation effects are provided at all internal additions and multiplications. A total of eleven coefficients are allowed for at present, which is sufficient for a third order filter with flexible scaling.

5. Conclusions:

This paper has described experimental equipment which has been constructed at the University of Sussex for the evaluation of two approaches to digital control of servomechanisms. Powerful and flexible facilities have been developed and are currently being used to make comparisons between incremental and integral transfer computers in this application.

6. Acknowledgement:

Acknowledgement is made to the Ministry of Defence for permission to publish this paper, and for supporting the work upon which it is based.

7. References:

1) Holtz,H., Leondes, C.T.,: The Synthesis of Recursive Digital Filters. Journal of the A.C.M. Vol.13, No.2. 1966. p. 262-280.

2) Braun, E.L.: Digital Computer Design. Academic Press. 1963.

3) Sizer, T.R.H.: The Digital Differential Analyser. Chapman Hall,1968.

NOTE:

This report describes work currently in progress at the University of Sussex. The results were not available at the time of printing but will be presented at the Conference.

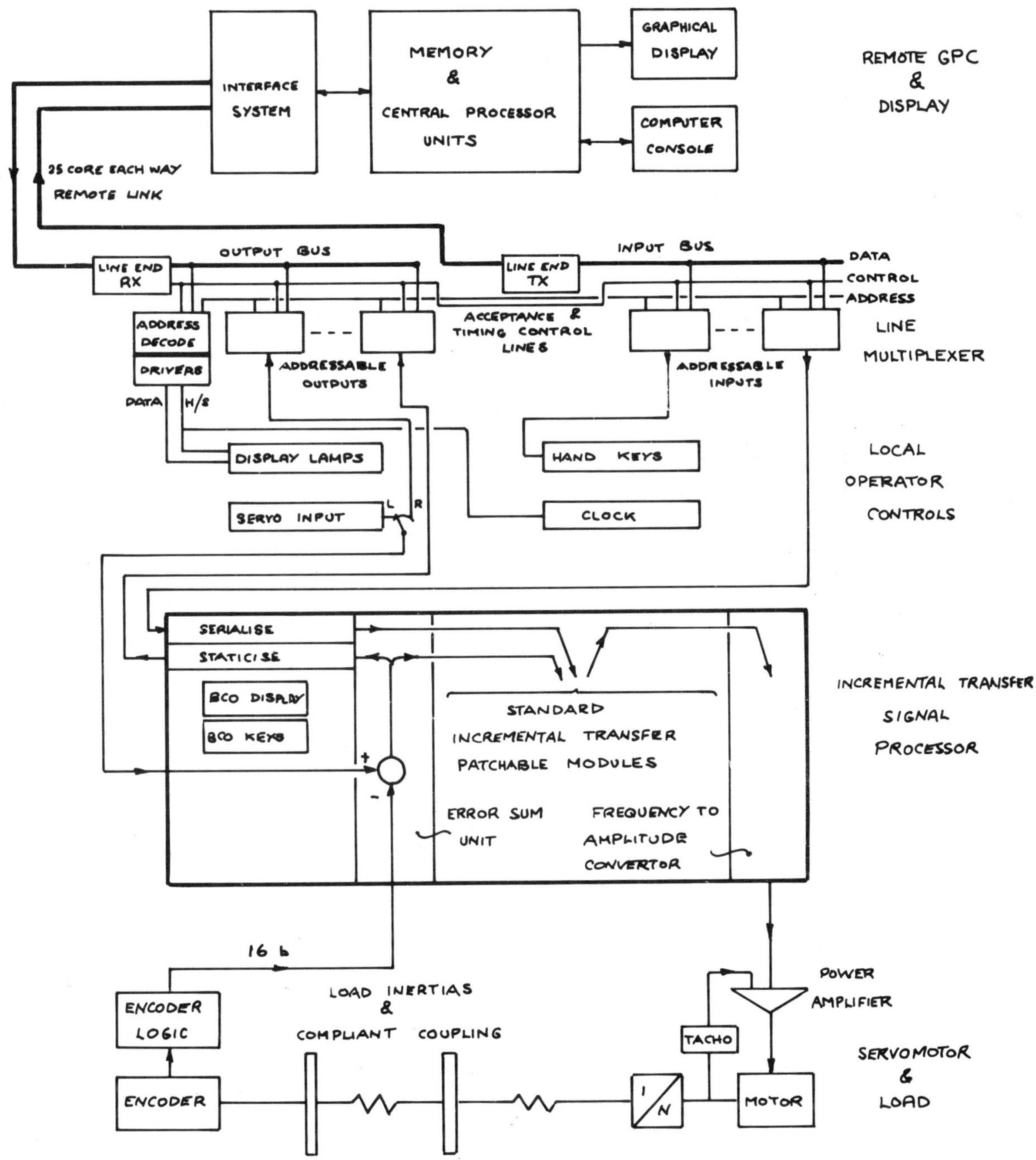

Figure 1 Overall System

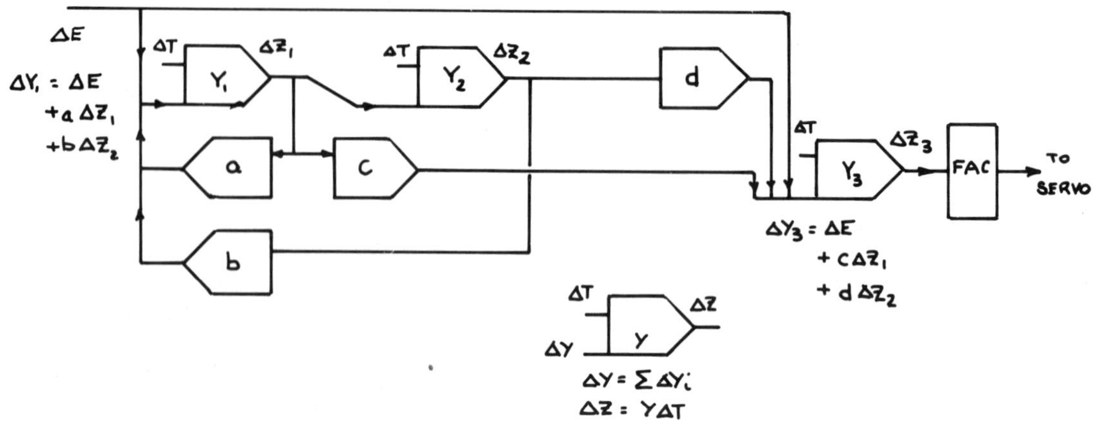

Figure 2 Typical compensator using Standard Modules.

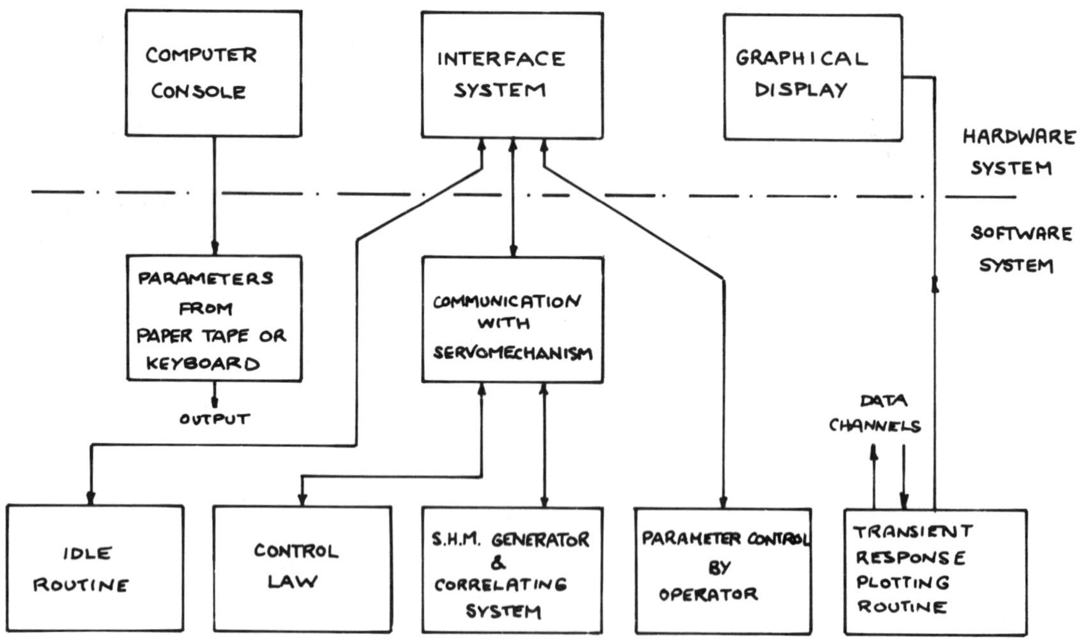

Figure 3 Program and Hardware Interconnection.

DIGITAL PANEL INSTRUMENTS IN SUBSTATIONS

D.J. Eastaugh and G.H.D. Keen

1. Introduction

Digital instruments have a considerable advantage over traditional analogue instruments in their clarity of presentation and this feature can usefully be employed for improving instrumentation schemes in transmission substations. Although this clarity is valuable in ensuring that accurate information can readily be obtained, the observation of trends is not so clear and consequently present practice is to employ both analogue and digital displays for appropriate duties in substation control schemes. This paper describes those schemes coming into use at new transmission stations at the present time. At a few stations associated with large new power stations, the quantities are displayed on Visual Display Units rather than on separate panel instruments to fit in with the overall station instrumentation scheme.

2. Advantages of Digital Instruments

The principle advantage of digital instruments over analogue types for substation control panels is their ability to give a clear reading that does not require any interpretation or decision on the part of the reader.

In times past, each primary circuit on a control panel was equipped with analogue instruments for indicating power, reactive power, current and voltage. The need to miniaturize control boards and to reduce the cost of equipment resulted in a change to a common set of instruments that could be switched to each circuit in turn. If all the circuits controlled from one control board were of similar rating then the single set of instruments could give adequate readings, but such was rarely the case. To preserve reasonable reading accuracy, multiple scales were adopted which required the operator to decide which scale he should be reading before attempting the task of interpreting the pointer indication. The possibility of error in such a system is large, particularly if the operator concerned is under stress as may be the case at times of system disturbance. It is at such times that it is essential to make a correct assessment of the situation and the consequences of acting on a wrong reading can be very serious.

By suitable adjustment to scaling factors and insertion of a decimal point, a single digital instrument can provide clear readings which require no decision or interpretation. Further, by using d.c. analogue signals the same type of instrument can be used for all the different quantities to be indicated; this has very considerable advantages from a maintenance point of view.

D.J. Eastaugh and G.H.D. Keen are with Central Electricity Generating Board, Transmission Division.

3. Disadvantages of Digital Instruments

Although superior for accurate reading, the digital type of display is not ideal for displaying a trend. A reader can deduce the trend in a reading from an analogue instrument by remembering, almost subconsciously, where the pointer was when he last looked at it. Further a cheap analogue instrument can show readily whether a circuit is "quarter full" or "full" or even "over full" by the position of the pointer, whereas with a digital display it is necessary to know the numerical value of "full" and perform a mental claculation.

From a maintenance point of view, digital instruments have considerable disadvantages. An analogue instrument can have a life of 50 years, with say 10 yearly checks to preserve performance adequate for switch board use. The maintenance requirements for digital instruments are certain to prove greater and there is some doubt as to the life expectancy. The digital instrument market appears to be very fluid, with new models appearing at frequent intervals and old models being abandoned. It is consequently very difficult to ensure that suitable replacements will be available during the life of a substation control board from both a technical and dimensional point of view.

4. Display Characteristics

None of the digital instruments at present available has a fully satisfactory display for substation use at an economic cost.

The cold cathode tubes suffer deterioration with age and are not satisfactory in high ambient light conditions. The seven bar digits can give wrong readings on the failure of a bar and need time for an operator to get used to the shape of the figures. The matrix display shows the greatest promise but at present the cost of such devices is unacceptably high.

Choice of display time is important if the operator is not to be mesmerised by continually changing digits. In practice the use of cold cathode tubes with an unblinking display updated every 2 to 3 seconds has been found to give the most acceptable display; this gives the operator time to establish a reading before it changes. As the maximum reading required is of the order of 5000 it has been found possible to use a four figure display comprising three figures counting to 999 with the right hand figure fixed at zero. A "x 10" switch is provided which moves the reading one place to the left and lights a decimal point to the left of the fixed zero. It has proved difficult to obtain a production model which could be modified to provide the requirement for a "decimalised" instrument with a fixed last figure and slow display time.

One of the difficulties in obtaining equipment suitable for this type of application arises from the natural reluctance of instrument manufacturers to produce special instruments in relatively small numbers.

5. Application

5.1 Power, Reactive Power, Current and Voltage

For displaying readings of all these quantities a standard 0.9990 volt instrument is used. Transducers with a nominal 10 mA d.c. output, reversible for power and reactive power, are connected to the primary current and voltage transformers. The output of each transducer is connected across a scaling resistor of such value that 1 mV is developed across the resistor for each unit of the primary quantity irrespective of the circuit rating or input/output characteristics of the transducers. The outputs of these scaling resistors are then selected onto common digital instruments of which there is one each for power, reactive power and current. For voltage indications one instrument is provided for each busbar section, and although it is not strictly necessary to display the fixed zero in this application there are advantages in using the standard instrument.

In addition to the digital instruments described above, which are intended to allow an accurate assessment of circuit conditions, small analogue instruments, for current measurement only, are provided on the control board itself. These are mounted in the mimic diagram so that the operator can observe system trends, particularly necessary when he is carrying out switching operations.

The only other analogue instruments used are those associated with synchronising where it is considered advantageous to display incoming and running voltages in this form to facilitate a correct comparison.

5.2 Frequency

The display of frequency can best be achieved by use of counting techniques rather than analogue to digital conversion. An instrument, specially developed for this purpose and using a time interval conversion method, has proved very successful. A five figure display with a resolution of 0.005 Hz and a display time of about 2 seconds is employed.

5.3 Transformer Tap Position and Winding Temperature

The same type of arrangement as that used for display of the system electrical quantities is also used for displaying transformer tap position and winding temperature. In the case of tap position, a resistance transducer is used to give an output proportional to the resistance of a tapped bank of resistors, the tapping points corresponding to the position of the main tap changer. The transducer output signal is then connected to a scaling resistor and the resulting voltage applied to a digital instrument similar to those described above but with display suitably modified to show up to 19 tap positions.

The present practice for measuring winding hot spot temperature is by means of a dial thermometer with a capillary connection to a heat sensing bulb mounted in the transformer tank and compensated for load current. A variable resistor is connected to the indicating device and this is used in conjunction with a transducer to give remote digital indication of winding temperature.

6. Interference and Power Supplies

It is clearly necessary for the instrumentation in transmission substations to remain unaffected by disturbances due to system faults. The problem can be divided into two parts, the first relating to power supplies and the second relating to possible damage due to transient over-voltages. Both these problems need careful consideration in the application of digital instruments.

Most digital instruments require a 250 V a.c. supply and all the no-break supplies at present provided in substations are obtained directly from batteries. It is necessary therefore, for those applications where continuity of indication is essential, to provide an inverter.

Provisions for guarding against damage due to high transient over-voltages must also be built into the system by using suitable cabling and ensuring adequate screening of the various components associated with the instrumentation scheme.

7. Conclusion

Although the digital instrument, despite its high initial cost and maintenance requirements, can prove an economical proposition there are certain factors that, from the supply utilities point of view, need to be borne in mind.

a) There is an expanding market available for a digital instrument that is tailored to the concepts of "switch board" instruments in respect of reliability, life, availability of spares and continuity of availability of replacement.

b) The question of power supplies needs careful consideration to ensure that the instrumentation is available at times of supply interruption, when it is most essential.

c) For switch board or control panel instruments, replacements should fit the same panel cut out. In this respect the DIN standards, or the British Standards at present under discussion could prove useful.

d) There is a need for a display that can be read in high ambient light conditions, is intelligible to a non-technical reader and that, when it fails, fails completely and not in a manner that can cause errors in reading.

A P.C.M. DIGITAL DATA ACQUISITION SYSTEM

G.A. Cooper

SUMMARY

This paper outlines the problems associated with high density data acquisition in an airborne environment, and describes the application of digital techniques to their solution.

Acknowledgment is made to The Plessey Company for permission to publish this paper, to the Ministry of Defence on whose behalf development has been carried out, and to I. & R. Dept., RAE, Bramshot, with whom a joint patent application has been made to protect the invention of the data formatting system.

1. INTRODUCTION

The continuing development of high performance modern aircraft is necessarily a continuous process relying heavily upon the feedback of accurate information with which to compare actual and predicted performance under a wide range of operating conditions. Information originates from an assortment of sensors, strategically placed throughout an aircraft. Sensors are of many different types, pressure transducers, strain gauges, synchros, etc., and a distinct problem of data gathering is that of converting these different signals into a standard form suitable for recording, without degrading their accuracy, either during the conversion process itself or by noise interference due to stray pick-up.

The bandwidths of the majority of sensors used on an aircraft is less than 5 Hz, but in the case of vibration measurements may be as high as several KHz. Signal levels vary from less than 10 mV (in the case of strain gauges) up to $\pm$5V (instrument potentiometers). The number of such parameters to be recorded may vary from say 20 to 30 for a small installation, up to several hundred where direct recording would be prohibitive in terms of weight and volume.

In view of the wide range of signal types, and input levels, differing bandwidths, and numbers of parameters to be sampled, to design a custom built system for each installation would carry a heavy cost penalty. To keep costs low therefore and yet offer flexibility a modular approach has been developed which provides a range of standard modules, each of which can be tailored very simply to meet the needs of the user. The temperature range of such modules when packaged in ATR boxes is -40°C to $+70^{o}$C.

G.A. Cooper is with the Plessey Company Ltd.

In order that the maximum amount of data can be recorded accurately P.C.M. digital recording techniques are employed. Thus by multiplexing and the use of variable sampling rates, higher bandwidth signals can be recorded at high sampling frequencies and lower bandwidth signals at appropriately lower frequencies and system efficiency is high.

Since the recording is in digital form it can be readily interfaced with a computer to give swift, automatic analyses.

2. ANALOGUE SIGNAL CONDITIONING

Analogue input signals range from $\pm$10 mV to $\pm$5V and originate from sensors located at some distance from the signal conditioners. Since these sensors are likely to be bonded to the airframe, a potential difference may exist between sensor and conditioner, due to earth currents flowing through the airframe. This voltage may distort or swamp the true sensor output. Therefore in this system all analogue inputs are considered as 2 wire differential inputs during both conditioning and multiplexing, and are only changed to single ended for conversion by the A/D converter.

To obtain 12 bit resolution and maintain accuracy, the A/D converter requires input levels with a fullscale range of 10V. To provide such a range from low level inputs, a standard differential to differential conditioning module has been developed. The gain of this module can be varied from 1 to 500 to suit the sensor and common mode rejection ratio is 80 dB at less than 1 KHz. In addition active or passive filters are available as standard low pass elements with cut off frequencies of 50 Hz, 10 Hz and 5 Hz.

3. MULTIPLEXING OF ANALOGUE SIGNALS

The standard module developed for analogue multiplexing is a 16-way, 2 wire M.O.S. package capable of accepting signals of up to $\pm$5V, plus a common mode voltage of $\pm$5V. The multiplexing element is a custom built device with very low leakage and a high 'off' impedance of 10^9 MΩ over the temperature range -40 to +85°C. The timing of the device gives the break before make action necessary to avoid cross coupling of sensors, and a capability of switching at speeds in excess of 32,000 samples/second.

4. SAMPLE AND HOLD

To reduce aperature time, and increase the operating speed of the system, when the settling time of multiplexers and buffers is greater than the acquisition time of a sample and hold circuit (i.e. when speeds greater than 16,000 samples per second are used) multiplexed signals are buffered by high speed operational amplifiers, sampled, then held and presented to the A/D converter for one system output word time. The buffer is necessary on this module to cater for the case where ±5V differential signals from a sensor are driven directly into the multiplexer, and provides good common mode rejection and high input impedance. Two output word times are required to convert an analogue signal into digital data. During the first word the multiplex switches are opened and the analogue signal is first sampled, then held. During the 2nd word, the held signal is converted into a digital number by the A/D converter, and at the end of the word time presented to the internal data highway.

In the case of unipolar OV to 5V signals, the analogue buffer and sample/hold board multiplies the signal by 2 and offsets it, to present a signal to the A/D converter of between -5 to +5V. An input signal change of OV to +5V unipolar, or -5V to +5V bipolar, is indicated by a number change of 4000 from the A/D converter.

5. ANALOGUE TO DIGITAL CONVERSION

A unique feature of the A/D converter module is its ability to correct for internal offsets due to drift with temperature change. This is achieved by dedicating one channel of a multiplexer to a calibration module.

The calibration module simulates a signal conditioner and is switched alternately between OV and a highly stable ¾ fullscale voltage under control of the sampling programme. When the programme calls up a OV output from the calibration module, an A/D conversion is performed in the normal way and a digital number representing OV is presented to the output highway. At the same time this digital number is read into a store on the A/D module, and, if greater than true zero is subtracted, if less than true zero is added to successive samples until updated at the next OV command from the programme.

Thus, when the ¾ fullscale voltage from the calibration module is selected, it is digitally corrected for drift, and may be used as a measure of system performance.

6. DIGITAL SIGNALS

6.1. Parallel

A module has been developed to accept two 12 bit parallel data words, multiplexed under programme control. The data words normally consist of flag or other 'EVENT' type signals and up to eight such multiplexing modules are fed into a digital buffer module which compares the input level on each wire with an adjustable threshold. The threshold is normally set to accept ARINC type 28V signals, where -1V to +3V is specified as digital zero and greater than +7V as digital one. Since this threshold is adjustable the standard module can be used to accept a range of different signal input levels.

6.2. Serial

To provide for simple expansion of the system an interface module has been designed to transmit serial addresses to a remote peripheral unit and receive appropriate serial data in return. This has the advantage that remote peripheral units can be placed in close proximity to the sensors they serve, thus reducing noise interference, and that digital data alone is transmitted over long cables to a central processing unit.

To reduce the possibility of induced noise degrading the data, all serial transmissions are made via two wire twisted pair screened cables with 5V differential drivers at one end and differential receivers at the other. The common mode rejection of these transmitter/receiver pairs is ±15V on a 5V signal, and each interface module can communicate with four remote units.

6.3. Documentary Data

The addition of a Control Unit to a data acquisition system has the following advantages -

(i) Identification of aircraft, flight, etc., can be recorded by setting in documentary data of from 0000 to 9999 on edge switches located on the front panel of the Control Unit.

(ii) Any one of 8 independent sampling programmes can be selected at any time merely by operating an edge switch.

(iii) EVENT markers can be inserted into the coded data stream and recorded on tape to facilitate data identification during recovery.

(iv) A continuous, non volatile record of tape footage remaining is displayed on the front panel.

(v) Tape recorder speed selection can be over-ridden to enable fast re-wind of recorded tape, and OFF/STANDBY/ON switching is provided to allow recorder warm-up at temperatures below $0^{o}C$.

(vi) Built in test equipment is available in order that by means of a dedicated edge switch, faults can be localised to one unit in the system.

7. DATA FORMATTING

Data formatting consists of generating a series of parameter addresses, one for each time slot. These addresses control the multiplexing switches to feed analogue signals to the A/D converter and digital information direct to the internal data highway.

7.1. General Considerations

The object of the P.C.M. data acquisition system is to sample data from a multitude of sources and at a frequency of sampling dependent upon the parameter. This multiplexed data must be arranged in such a way that recovery of information is both simple and reliable. As with telemetry systems, reconstitution of sampled data is greatly simplified if successive samples of a given parameter are equally spaced in time.

With these objects in mind the data format is restricted to an even number of time slots, each of which is termed word period, and represents a parameter sample. Samples of signal inputs occur at integer time slot multiples to satisfy the equal spacing constraint and are binary related. These samples are grouped together in Data Frames, and transmitted as a serial data stream.

A sequence of consecutive Frames is termed a Data Field and spans the period of the lowest sampling rate. To differentiate between Data Frames and Data Fields a unique synchronisation pattern called up by the programme is inserted in the data stream to indicate the beginning of a data field. The same pattern is inverted and inserted in the data stream, to indicate the second and subsequent data frames within the data field. This technique is termed 'Frame Code Complement' and is also employed to synchronise the demultiplexing routine necessary for data recovery. There are many approaches to the allocation of time slots to parameter samples, each with its own merits. With these in mind a new system has been developed to meet the following requirement.

(1) The method of programme storage must be non-volatile and occupy minimal volume.

(2) The programme must be capable of selecting parameters at as wide a range of sampling rates as possible.

(3) The data formatting must be automatic and programme writing simple to save laborious computer programming.

(4) Parameters must be sampled in an order dictated by their sampling rates and not by their location.

(5) Programme changes must be simple and flexible to cater for different trials.

(6) Recovery of recorded data must be as automatic as possible in order that demultiplexing software may be kept to a minimum.

Requirements (1) and (5) have been combined by the use of Programmable Read Only Memories (PROMS) as the storage medium. Previous data acquisition systems have used methods such as a diode matrix for storing a programme sequence but the volume required is prohibitive for a small system, and the number of such programmes stores is severely restricted. Using a single 256 x 4 PROM, up to 8 completely independent programmes are stored, any of which may be selected by a switch on the Control Unit. The advantage of such simple selection is that sampling rates of parameters can be immediately adjusted to suit different phases of a flight envelope. For example during automatic landings where 4 minutes of recording are required on average, sampling rates of autopilot responses can be increased above those of normal flight merely by selecting the appropriate programme.

Requirements (2) to (4) are satisfied by the programme generator which controls the following variables independently -

(a) Parameter sampling rate

(b) Data Field length

(c) System output word rate

(d) Tape recorder speed

7.2. Programme Generation

Generation consists of producing a series of parameter addresses, one for each time slot in the Data Field. In order that storage requirements are not dependent upon the length of the Data Field, the following system has been developed where addresses need only be stored once, but are called up as many times as a parameter is required to be sampled. Hence the volume of storage is dependent upon the total number of parameters alone, irrespective of sampling rates. Generation of programme has been divided into two distinct operations - Scanning and Translation.

7.2.1. Scanning

The scanner generates a pattern such that samples are evenly spaced in time. The output consists of groups of binary numbers, one group for each sampling rate, repeated at that sampling rate and interleaved. There are 7 binary related sampling rates per field: 64:1, 32:1, 16:1, 8:1, 4:1, 2:1, 1:1, and field lengths vary according to the number of samples at each rate.

At the highest sampling rate (64) for instance, a group of numbers will be repeated 64 times throughout the pattern at regularly spaced time slot intervals. At the next highest sampling rate (32), a second group of numbers, interleaved with those at rate 64, will be repeated 32 times throughout the pattern at regularly spaced intervals. At the next highest rate (16), a third group of numbers, interleaved with those at both rate 32 and rate 64 will be repeated 16 times throughout the pattern, again at regularly spaced intervals. This process continues for successively lower sampling rates down to once per field.

Each group of numbers is sequential and increments by one for every sample.

This pattern generation is fully automatic, the information required for a 256 parameter system is as follows -

(a) Two 8 bit words at each rate, one to denote the start and one for the finish of each group - giving 14 words.

(b) A single word to define the number of time slots between each repetition of the groups at the highest sampling rate, and hence the number of time slots in a data field.

(c) A single word with one half defining the system output word rate and the other half an appropriate tape recorder speed.

These 16 instructions are stored in a PROM in the form of 32 four bit words, thus 8 programmes can be stored in a 256 x 4 PROM, each of which can control sampling rates, system word rates, pattern length and recorder speed.

7.2.2. Translation

Since the start and finish of each group of numbers generated by the scanner is variable, parameters can be scanned sequentially at each rate, which for small systems is adequate, but for large systems would necessitate grouping together inputs according to their sampling rate.

To provide a simple method of digitally rewiring the inputs to suit different sampling patterns, a translator is used whereby the numbers from the scanner each call up a storage location. This location contains the address of the parameter which can be located at random within the system. The storage medium is again 256 x 4 PROMS, contained on a single Translator Store module, capable of storing 256 such addresses.

7.2.3. Programming Procedure

This requires no special training and a large number of variables can be controlled quite simply. The procedure consists of -

(1) Listing the parameters to be sampled and their ideal sampling frequency (normally five times maximum bandwidth ignoring aliasing errors).

(2) Assembling these parameters in groups at binary relating sampling frequencies, starting with the highest and working down, and allocating a relative rate to each frequency.

(3) Calculating the pattern length in time slots by summing the products of groups and rates.

(4) Selecting a system word rate from 10 binary related frequencies (32 to 32, 768 Hz) dictated by the highest sampling rate of the pattern.

(5) Selecting an appropriate recorder speed within the tape packing density capability.

Programme details are then read into the Programme Stores either automatically or manually.

8. DATA RECOVERY

This system possesses the unique feature that whenever -

(a) The tape recorder is started

(b) Power supplies are temporarily interrupted, or

(c) Programme is changed

a 'dump' sequence is initiated. The dump sequence records a normal data field with sync patterns, but with parameter addresses substituted for data. This process is preceded by fields of logic zeros followed by logic 1's and continues for 8 fields. Thus ground recovery equipment is supplied with the necessary demultiplexing information automatically, prior to each data run, and the need to reprogramme software each time programme is changed is obviated. This function is performed by a single optional Programme Dump module.

9. SYSTEM CONSIDERATIONS

Modules already described form the basis of present systems and can be packaged to suit the user. Standard packaging consists of ½ ATR boxes, both long and short, according to the conditioning, types, and number of signals per system. Additions to the basic range of modules are under consideration and future systems are expected to include adaptive processing and hence data reduction during flight.

ON-BOARD DIGITAL ANALYSIS OF LOW FREQUENCY RADIO SIGNALS ON SPACECRAFT

Dyfrig Jones

Introduction:

A study of radio signals at frequencies below 100 kHz observed on ground and on spacecraft can yield considerable information on the medium through which the waves have propagated. If the signals are of natural origin, they may cover a wide frequency range. "Hiss", for example, which has characteristics similar to band-limited random noise, may cover tens of kHz and is believed to be produced by wave-particle interactions in the ionosphere and magnetosphere. "Whistlers", on the other hand, are narrow-band phenomena whose frequency may sweep from above 30 kHz to below 1 kHz in a time of about 1 sec.; they have their origin in lightning discharges near the earth's surface, and the frequency dispersion is caused by propagation through the magnetosphere.

A detailed analysis of these and other events presents little difficulty on the ground where they may be recorded directly on magnetic tape or analysed in real time by specially designed spectrographs. However if the signals are to be investigated on spacecraft, a considerable amount of transmitter power is required to telemeter even a 10 kHz bandwidth to ground. Hence it is important to investigate the possibility of processing the signals on board the spacecraft so that the telemetry rate can be reduced. This has led to the inclusion of a digital correlator on the ESRO geostationary satellite GEOS which is to be launched in 1976. Such a correlator has a digital memory and this can be put to a number of uses relevant to particle measurements in space as well as to electromagnetic and plasma wave studies. The most important applications are summarised in this paper.

Analysis of Very Low Frequency (VLF) Signals:

It has been usual to study natural VLF phenomena by performing two major types of analysis. The first - auto-spectrographic analysis - produces a three dimensional display (spectrogram) having frequency as ordinate, time as abscissa and signal strength as intensity modulation. The second - cross -spectrographic analysis of two wide-band signals - yields information on their relative phases and their coherency. The coherency may be displayed by intensity modulation and the phase can be represented in colour[1] or vectorially[2].

Auto-spectrographic analysis. From the manner in which a natural VLF signal varies in frequency with time, one can obtain information regarding the medium of propagation. For example, spectrograms of whistlers observed on the ground have led to estimates of electron density extending to more than six earth radii from the earth, whereas spectrograms of other phenomena observed on spacecraft yield information on the local magnetic field, plasma density and energetic particle flux.

A range of apparatus for producing auto-spectrograms on the ground have envolved, extending from a narrow-band swept-filter analyser to a multi-filter analyser. The former has frequency and time resolutions of the order of 50 Hz and 20 ms respectively for an 8 kHz bandwidth, whereas the latter

D. Jones is with the European Space Research and Technology Centre, Noordwijk, The Netherlands.

may consist of as many as 512 filters, whose 50 Hz bandwidth overlap to cover a total band of 20 kHz, and whose outputs are sequentially sampled every 20 ms. Such analysis were extensively used for ground VLF studies and it was natural that their use was extended to signals recorded on spacecraft. For example, the NASA Imp series of magnetospheric spacecraft employed analogue telemetry links along which the signals were transmitted in raw form. On reception at ground the signals, which had bandwidths up to 30 kHz in some cases, were analysed by a conventional spectrograph. This procedure has a number of disadvantages, the most important being:

(a) Noise unavoidably enters the telemetry link, resulting in a degradation of the final spectrograms. This effect may be reduced by increasing the transmitter power.
(b) Special analogue facilities are required on board and on ground since most other data are telemetered in digital form.
(c) The signal can only be telemetered intermittently since the transmitter power required is usually too large to allow continuous coverage.
(d) The telemetry is being inefficiently used, bearing in mind the type of analysis to which the signal is subjected later. Auto-spectrographs disregard phase information and in the process of spectrum analysis there exists a limiting time resolution imposed by filter circuits. Hence, although one may determine the amplitude of the signal at each point in time, this information is in effect averaged out with a certain time constant in the analysis process.

Both (a) and (b) may be avoided if the signal is telemetered in digital form, and with the computer facilities now available there is little difficulty in producing digital spectrograms comparable in quality to their analogue counterparts. However, objection (c) is also valid for digitised signals. The ideal solution would be to perform the complete analysis on the spececraft and to transmit the spectra to ground in digital form. Fast Fourier Transform circuits are available and could be used for this purpose were it not for their heavy demands on the limited weight and power available.

A function which contains the same information as the power spectral function, but which is more amenable to its on-board computation, is the auto-correlogram. This discards the phase information, and the averaging time may be designed to be of the same order as would be imposed by a ground spectrograph. After this integration time, the correlogram would be telemetered to ground where it would be Fourier transformed to yield the power spectrum. In this way, only the essential information is transmitted, and the raw signal is replaced by a succession of correlograms which are used to form the spectrogram.

The GEOS digital correlator can accommodate 64 samples, and since the auto-correlation function is symmetrical, only one half of it is computed and transmitted[3]. The integration time is presently fixed at 30 ms and the frequency resolutions in the final spectra will be 50, 100 and 200 Hz for the bandwidths 2.5,5 and 10 kHz respectively. By heterodyning the signal, the 5 kHz bandwidth may be placed at any frequency up to 80 kHz. The signal is digitised to 4-bit accuracy and the correlogram is telemetered with 8-bit accuracy. The total telemetry bit-rate is therefore 17 kbits/sec, as compared with 100 kbits/sec required to telemeter a 10 kHz signal with 4-bit accuracy.

One consequence of performing on-board correlation analysis is that the signal cannot be reconstructed from the correlogram, and one may not re-analyse the signal with different frequency and time resolutions. On GEOS, this limitation is alleviated to a large extent by choosing one of three bandwidths by command.

Although the foregoing discussion has been centred on the analysis of natural VLF signals, a correlator of the type described may be used for the analysis of virtually any signal. For example, it is obvious that the flux of low energy electrons in the magnetosphere is modulated by the presence of an alternating electric or magnetic field. Therefore if the particle flux were sampled at a certain rate and the auto-correlation function formed, one should arrive at an estimate of the power spectra of the alternating fields.

Cross-spectrographic analysis. In a number of cases it is important to determine the coherency and relative phases between two or more wide-band signals. To describe completely an electromagnetic wave in space requires the measurement of its three electric components $E_{x,y,z}$ and its three magnetic components $E_{x,y,z}$ where x,y,z refer to the spacecraft reference-frame. By computing the cross-spectral functions for various combinations of the components, information on the wave structure may be extracted. Some cases are considered briefly here in ascending order of complexity:

(i) With spacecraft spin axis z parallel to magnetic field, the sign of cross-correlogram E_xH_y yields an estimate of Poynting flux direction.[4]

(ii) In addition to electromagnetic waves it is possible for electrostatic waves in the same frequency range to propagate in a plasma, and if only the output of the electric antenna is examined, one cannot distinguish between them. However, if the outputs of one electric and one magnetic antenna are cross-correlated, weak electromagnetic signals may manifest themselves since their electric and magnetic components are correlated in contrast to the electrostatic waves which have no magnetic component.[5]

(iii) A large number of natural VLF phenomena have distinctive polarisations; for example, a whistler is right-hand circularly polarised when it traverses the magnetosphere so that if it were received on two orthogonal antennae in the plane of the wavefront, the phase difference between the signals would be $\pi/2$. This phase signature may be used as a means of recognising low amplitude phenomena in the midst of noise.[6]

(iv) Simultaneous computation of the three cross-correlation functions between the magnetic components $H_{x,y,z}$ of an electromagnetic wave will yield the direction of the wave-normal.[7]

(v) Accurate determination of the Poynting vector direction requires six cross-correlation functions E_yH_z, E_zH_y, E_zH_x, E_xH_z, E_xH_y and E_yH_x.[8]

(vi) Wave polarity may be studied from 6 auto and 15 cross-correlograms.[9]

In the past those of the above computations which have been performed on wide-band signals were done on the ground. The required wave components were telemetered in analogue form and cross-correlated after reception. It follows that for a given telemetry capacity the more the number of components required, the less the bandwidth which may be ascribed to each. In addition to the disadvantages ascribed to analogue telemetry which have been described, one has the phase inaccuracies in the link and recording systems. Higher phase accuracy is achieved with digital telemetry, but the bit-rate increases directly as the number of channels increases; objection (c) is also valid.

The digital correlator on GEOS will be capable of computing the cross-correlation function of two signals and its operation is illustrated in Fig. 1. The signals are digitised to 4-bit accuracy and after an integration time of 30 ms, the correlograms are telemetered with 8-bit accuracy. The same bandwidths will be covered as for auto-correlation, but the frequency resolution in the cross-spectrum is decreased by a factor of 2, since the cross-correlogram is an odd function which requires the computation of both positive and negative portions.

An objection to telemetering only the cross-correlogram is that signals which are uncorrelated will be lost. This may be overcome if the auto-correlograms are also computed either simultaneously or alternately as will be the case on GEOS. A detailed description of the operation of the GEOS correlator may be found in reference 3, and it allows the investigation of (i), (ii) and (iii).

An on-board digital correlator proposed for the Imp KK' (Mother/Daughter) spacecraft to be flown in 1977 was designed to compute simultaneously three cross-correlation functions.[10] This would yield the wave-normal direction for 16 frequencies uniformly covering a 4 kHz bandwidth placed anywhere between zero and 256 kHz, and hence would allow the fulfilment of (iv) in addition to (i), (ii) and (iii). One can envisage adapting and extending such a correlator to cope with the 6 cross-correlograms required by (v). However, there is a more efficient method of dealing with this case, which is also applicable to (vi). Before discussing this method, the application of on-board correlation analysis to signals produced by active experiments will be considered briefly.

Correlators for Active Experiments:

Consideration has been given to the reception and on-board analysis of natural VLF signals. However, an on-board correlator may also be used for active experiments in which the signals generated on ground or in space are received on a spacecraft. Two examples will be given - a plasma resonance experiment on GEOS and an experiment on Imp KK' where signals are transmitted from the Mother to the Daughter spacecraft.

Plasma resonance experiment. When an antenna is excited by a transmitter at one of the characteristic frequencies of the plasma in which it is immersed, a pronounced "ringing" at this frequency or its harmonics is observed and this forms a powerful technique for probing the ambient plasma.[11] An experiment to study these resonances must explore systematically a band of frequencies within which the resonances are expected to occur. Owing to telemetry limitations, it is not possible to investigate simultaneously the whole band of interest which is zero to 80 kHz for GEOS. The available telemetry is efficiently used if the signals are heterodyned to within the bandwidth of the correlator. Spectrograms are produced on ground as for natural VLF signals and from these one can monitor the state of the plasma.

Propagation experiment. The Imp Mother spacecraft will be capable of transmitting signals to the Daughter for the purpose of investigating the plasma between them.[10] The orbits are such that the spacecraft will be approximately 1000 km apart with their apogee in the solar wind at a distance of 23 Earth radii. The signals will consist of a 1 MHz C.W. transmission and a 300 MHz carrier modulated at 1 MHz; The modulation and the 1 MHz C.W. are in phase on transmission, but due to the effect of the plasma on the 1 MHz signal, they will not be in phase on reception at the Daughter. After demodulation of the 300 MHz signal, the phase difference between the 1 MHz waveforms yields an estimate of the integrated electron density between the spacecraft. To perform the phase comparison digitally is a relatively simple matter since one is dealing with monochromatic signals at a known frequency.

Using the principle of aliasing, the 1 MHz signals may be transformed to lower frequencies at which on-board digital computations are not as power-demanding. Consider a high frequency f_0 which is sampled at a lower frequency f_s such that $f_s = 2f_0/(n+\frac{1}{2})$. Then the signals f_0 will be aliased down to $f_s/4$ i.e. to half the Nyquist frequency. If $f_0 = 1$ MHz, $n = 100$, then $f_s = 19.9$ kHz and f_0 will be aliased to 4.975 kHz. But digitising a monochromatic signal at the rate of 4 samples per period results in samples $\pi/2$ apart. Consequently, if the two 1 MHz signals are simultaneously digitised, the cross-correlogram will also have samples $\pi/2$ apart. It is well-known that the phase of a sinusoid may be determined from two such adjacent samples. Therefore the phase difference between the two 1 MHz signals may be obtained from a 2-word correlator as illustrated in Fig. 2. The two words R0 and R1 correspond to a lag of zero and $1/f_s$ respectively. After a certain integration time, e.g. 100 ms, the two words are telemetered to ground and the phase shift obtained directly:(phase of signal 1 - phase of signal 2) - arctan(R0/R1).

Rotating Memories

It has been indicated that a thorough investigation of six wide-band electromagnetic components as is required for (v) and (vi) is not practicable using on-board correlators and it is not efficient to transmit the signals to ground in raw form. In this case it is useful to consider rotating memories as a means of performing on-board analysis. This analysis may also be applied to long-baseline interferometry on multiple spacecraft and on the ground.

Consider two wide-band signals S_1 and S_2, whose highest frequency is f_N, to be simultaneously digitised at frequency $2f_N$. If the cross-correlation function is formed with maximum time-lag of $\pm L/2f_N$, then on Fourier transformation, spectral estimates will occur at nf_N/L where n = 1,2....L. Consequently, if S_2 is a known signal consisting of equal amplitude sinusoids with predetermined phases at these L frequencies i.e.

$S_2 = \sum_{n=1}^{L} s \sin(2\pi f_N t/L + \emptyset_n)$ where $t=m/2f_N$ and m=0,1,2....., and all $\emptyset_n$ are known, then from the cross-spectrum the power and phases of the corresponding frequencies in S_1 are obtained. The waveform S_2 could be generated on the spacecraft and could be correlated with any unknown signal received. Signal S_2 is repetitive with period $T=L/f_N$ and as it is sampled at frequency $2f_N$, only 2L samples are required to describe the waveform completely. Therefore if one had a digital memory containing these 2L samples and they undergo one complete rotation in time T, the effect is to produce the waveform S_2. Correlation with S_1 may then be performed as illustrated in Fig. 3, in which each incoming sample of S_1 is multiplied by all 2L samples of S_2 and the product added into the stationary correlation register. By the time the next S_1 sample arrives, all S2 samples will shift by one location. After a suitable integration time, the correlation function is transformed to the buffer prior to telemetering, thus leaving the correlation register free for the next computation i.e. it is a real-time process. The rotation of the S_2 register, and the sampling of the ADC for S_1, are synchronised to the spacecraft clock. Hence, after Fourier transformation of the correlation function, the phases of the S_1 component frequencies will be known with respect to the spacecraft clock.

It is interesting to consider the form of signal S_2 for various values of $\emptyset_n$; three examples are shown in Fig. 4. The upper waveform is for the case where $\emptyset_n$ are random; the second is for $\emptyset_n=0$ and the third is for $\emptyset_n=(n-1)3\pi/2$. The power spectra of all three waveforms are identical. An interesting point which emerges is that if the third is digitised to 4-bit accuracy, the waveform approximates to the binary number N at time $pT+L/4f_N$ where p=0,1,2,...., and all other samples of the waveform are zero. Therefore with reference to Fig. 4, the S_2 memory may be discarded since all locations except one would contain zeros, and the cross-correlogram is formed by multiplying each S_1 sample by the number N and adding into the appropriate location in the correlator register. But N is then a mere scaling factor which could be replaced by unity. Hence if the correlation store is rotated at a rate of one complete revolution in time T i.e. one location per sampling interval, and the S_1 samples are added as they arrive into a fixed location of the correlator memory, the result is, in effect, a cross-correlation with the third waveform of Fig. 4. It is obvious that the process is merely an averaging of the signal S_1 and after a suitable integration time, the correlator register will contain only those frequencies which have added in phase i.e. the frequencies nf_N/L. All other frequencies, whether noise or signal, will tend to average to zero and the result is equivalent to passing the wide-band signal through L narrow-band filters at frequencies nf_N/L. The equivalent filter bandwidth decreases as the averaging time is increased, and it must be assumed that the signal S_1 is statistically stationary during integration.

Six-component electromagnetic wave measurements. The bit-rate required to telemeter directly six components each of 500 Hz bandwidth with 4-bit accuracy is 24 kbits/sec. Using six rotating memories each of 16 locations, and an integration time of 100 ms gives a bit-rate of 7.68 kbits/sec if the contents are telemetered with 8-bit accuracy. These six functions are then Fourier transformed to yield estimates of the power and phase with respect to the spacecraft clock, at 8 frequencies spaced 62.5 Hz apart in the range 0-500 Hz. At these 8 frequencies, computations of the type indicated in (v) and (vi) may then be performed on ground.

Long-baseline interferometry (LBI). In an effort to achieve higher resolution by extending the baseline and avoiding ionospheric effects, LBI is presently being considered using two spacecraft. When the problems concerning varying baseline length and direction have been solved the problem of data acquisition remains. It may not be feasible to transmit directly from one spacecraft to another and hence there arises once more the question of telemetering wide bandwidths with high phase accuracy to ground. A solution is to have a rotating memory on each spacecraft so that after an optimum integration time, the averaged waveform is telemetered and the phases of the various frequency components will be known relative to accurate spacecraft clocks. The memories should be capable of accommodating 256 samples,[12] allowing phase comparison at 128 frequencies within a chosen bandwidth. Similar rotating memories may be used for ground LBI if one wishes to reduce the amount of data to be recorded.

Acknowledgement. The author wishes to thank K. Strohm for assistance with the computations.

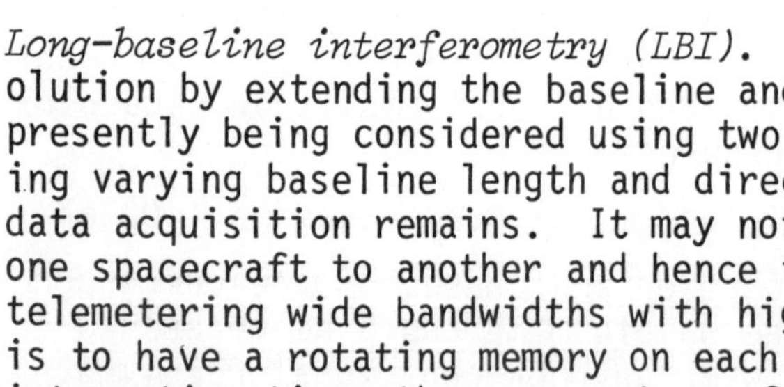

Fig. 1:

GEOS on-board digital correlator: configuration for cross-correlation.

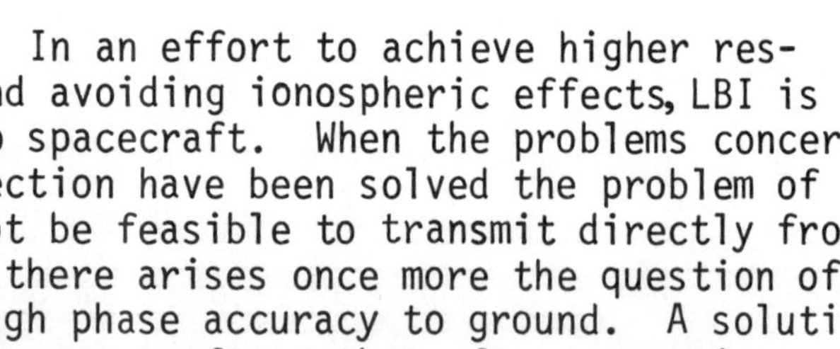

Fig. 2:

Phase-measuring 2-word digital correlator for monochromatic signals.

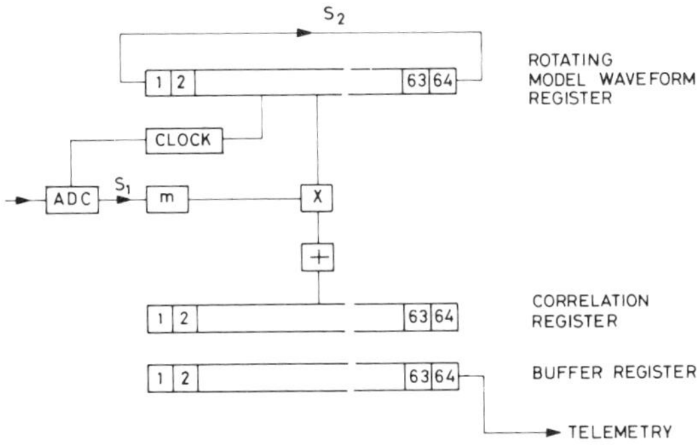

Fig 3:
Correlating a known signal S_2 stored in a rotating memory, with an unknown signal S_1.

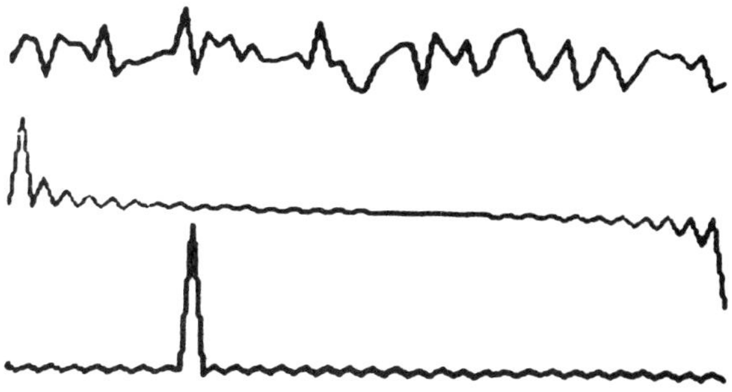

Fig 4:
Waveform $S_2 = \sum_{n=1}^{L} s \sin(2\pi f_N t/L + \emptyset_n)$
where $t = m/2f_N$ and $m = 0, 1, \ldots 2L$
with $s = 1$, $L = 32$, and (a) $\emptyset_n$ random
(b) $\emptyset_n = 0$, (c) $\emptyset_n = (n-1)3\pi/2$.

References

1. D.A. Gurnett, S.R. Mosier and R.R. Anderson, "Colour Spectrograms of VLF Poynting Flux Data", J. Geophys. Res., Vol. 76, pp. 3022-0030.

2. D. Jones, "The Application of Cross-Spectral Analysis to VLF Signals", ESRO SN-123 (ESTEC), European Space Research and Technology Centre, Noordwijk, The Netherlands, 1973.

3. D. Jones, "An On-Board Digital Correlator for Spacecraft VLF Radio Wave Studies", IEEE Trans. Geosc. Electr., to be published, 1973.

4. S.R. Mosier and D.A. Gurnett, "Theory of the Injun 5 Very Low Frequency Poynting Flux Measurements", J. Geophys. Res. Vol. 76, pp. 972-977, 1971.

5. F.L. Scarf, R.W. Fredricks and G.M. Crook, "Detection of Electromagnetic and Electrostatic Waves on OV3-3", J. Geophys. Res. Vol. 73, pp. 1723-1746, 1968.

6. D. Jones, "Pattern Recognition for VLF Studies", Radio Science, Vol. 8, pp. 467-473, 1973.

7. J.D. Means, "Use of the Three Dimensional Covariance Matrix in Analysing the Polarisation Properties of Plane Waves", J. Geophys. Res., Vol. 77, pp. 5551-5559, 1972.

8. R. Grard, "Intrepretation à mesures de champ électromagnétique T.B.F. dans la magnétospherè, Ann. Géophys, Vol. 24, pp. 955-71, 1968.

9. L.R.O. Storey and F. Lefeuvre, "Theory for the Interpretation of Measurements of the Six Components of a Random Electromagnetic Wave Field in Space", Presented at the Sixteenth COSPAR meeting, Kontstanz, Germany, May 1973.

10. C.C. Harvey, M. Petit, J.R. McAfee, R. Grard, D. Jones, J. Etcheto and R. Gendrin, "An Investigation of Wave Fields and Plasma in the Magnetosphere and Solar Wind", Experiment proposal Imp KK', submitted to ESRO, Sept. 15, 1972.

11. M. Petit, "Transient responses: Resonance Spikes of Topside Sounders: Summary Report", Radio Science, Vol. 7, pp. 791-794, 1972.

DIGITALLY CONTROLLED RADIO FREQUENCY INSTRUMENTATION

P. SAMPLE

Introduction

With the growing interest in automatic test methods for radio frequency measurements considerable demands are being put on the test equipment manufacture to produce suitable programmable instrumentation. The performance of a radio frequency test system is highly dependent on the individual instruments which provide the stimulus and measurement functions for the equipment under test. To date there has been a tendancy to produce instruments with capabilities aimed at either a particular function or to suit a limited range of user requirements. For example the basic stimulus in a radio frequency test system is the frequency synthesizer. To design a synthesizer that functions simply as a single signal source with possibly limited modulation facilities can be uneconomic when used in an automatic test system. Figure 1 illustrates this point, by taking four commonly occuring R. F. measurements. The simplest possible configuration is shown in Figure 1 [a] where a single synthesizer is used to measure the gain of the system under test [S. U. T]. However, the other examples given in Figures 1 [b], [c] and [d] require a much more expensive approach in that two synthesizers are required. The significant feature of all three measurements involving two signal sources is that the frequency difference is usually small and is held constant for a series of measurements.

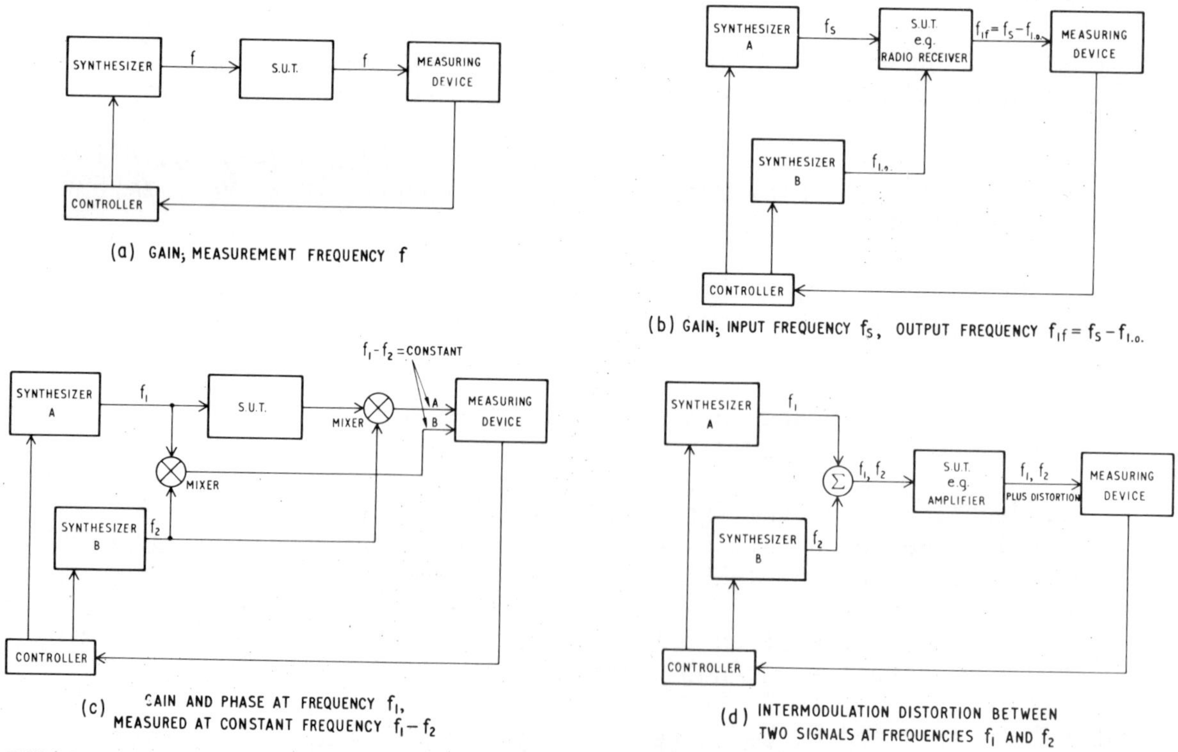

FIGURE 1. EXAMPLES OF R.F. MEASUREMENTS MADE WITH DIGITALLY CONTROLLED SYNTHESIZERS

P. Sample is Chief Engineer, R. F. Instruments, with Solartron - Schlumberger Limited.

This paper describes the design of a range of radio frequency instrumentation, one feature of which is to offer a much more cost effective solution to the two signal source measurements mentioned above. The basic instrument of the range is a 180MHz frequency synthesizer which has a very wide range of modulation and level control facilities and has all functions digitally programmable. A simple modular approach is used in the construction of the frequency synthesizer to allow a wide variation in the choice of functions. The frequency synthesizer may be easily converted to provide two output signals which track by a defined offset frequency by the addition of a simple add on unit.

The design philosophy of the frequency synthesizer is described in the following sections and the paper is concluded by a brief description of an experimental gain and phase measurement system using the frequency synthesizer with a digital transfer function analyser.

180MHz Frequency Synthesizer Design

Fundamentals

The frequency performance of any synthesizer is determined by the frequency standard which drives it. The various contributions to this performance may be listed as follows:

[a] Accuracy of output frequency
[b] Long term stability
[c] Short term stability, which includes phase noise contributions.

The reasons for the dependence of [a], [b] and [c] on the frequency standard are straightforward. Whatever the method of frequency synthesis the various blocks in the synthesizer provide simple mathematical operations on the output frequency from the standard to produce the desired frequency from the synthesizer. Any fluctionations in the frequency of the standard therefore cause corresponding fluctuations of the synthesizer output frequency. Long term variations in frequency are translated as a fixed ratio e.g. if the standard varies 1 part in 10^9 then the synthesizer output frequency will vary by 1 in 10^9. In the case of phase fluctuations at rates of several hertz and above the relationship is not so simple. For an ideal synthesizer if the instantaneous phase shift due to noise effects on the reference frequency, fr, is ϕ then the phase shift at the output frequency, fo, is given by

$$\phi o = \phi \frac{fo}{fr} \quad \text{-------------} \quad [1]$$

In a practical situation, however, to achieve the synthesis of the desired frequency a large number of frequency processing operations including multiplication, may be necessary which will modify the level of phase noise considerably.

For example in a synthesizer having four decades of frequency the following relationship could apply:

$$fo = fr\,[1 + 0.1N_1 + 0.01N_2 + 0.001N_3 + 0.0001N_4] \text{------} [2]$$

where N_1 to N_4 are integers between 0 and 9 and set the four respective decades.

Ideally equation [1] would apply in this example and in the worst case $\phi o = 1.999\phi_r$. However, in a practical situation it is not convenient to produce a system which would synthesise equation [2] directly. In order that frequency ranges covered by filters or phased lock loops may be fractionally small a more likely equation which can be realised in practice is given by:

$$fo = fr\,[1 + 0.1n_1 + 0.01n_2 + 0.001n_3 + 0.0001n_4] - 2.222fr \qquad [3]$$

where n_1 to n_4 are integers between 20 and 29 and set the four respective decades.

The first section of equation [3] indicates that the frequency processing is carried out at a much higher frequency with higher multiplication ratios and therefore the fixed offset frequency of 2. 222 fr has to be subtracted to obtain the original range of fo as given by equation [2].

Again in the ideal situation, assuming that the processing circuits are non dispersive and do not destroy the coherence of any multiplied phase perturbations, the effect of extra multiplication cancels when the various terms of equation [3] are combined. However in practice this will not be the case unless considerable care is taken in the design of the synthesizer. The result will be that coherence between the various multiplied phase noise contributions will be lost and when the output frequency fo is produced these will add to give a much higher level of phase noise then expected.

The above points can be summarised to give some very simple ground rules which apply to the design of any frequency synthesizer where overall phase noise performance is of prime importance:

[a] The frequency standard must have a very clean output. For example the typical phase noise density of a frequency standard for synthesizer applications is in the region of -170dB/Hz at rates corresponding to a frequency of 20KHz.

[b] Minimise the order of any frequency multiplication.

[c] Minimise the number of times that the reference frequency is processed.

[d] Maintain bandwidths, particularly of phase lock loops, as wide as possible if cancellation of coherent noise sidebands is required. As an exception to this rule, in some instances it is possible to produce secondary frequency standards in the synthesizer by phase locking a voltage controlled crystal or LC oscillator to some multiple of the main frequency standard. A narrow bandwidth phase lock loop would be used to filter the multiplied noise. This technique may be used only when the resultant noise level is below the limitations set by the above rules. The following sections show how the above basic considerations have been applied to the design of a 180MHz frequency synthesizer.

Requirements of the Frequency Synthesizer

The general design requirements can be listed as follows:

[a] Low noise performance; a threshold better than -140db/Hz.

[b] Fast programming of all functions; any frequency change is less than 1 m. sec. for example.

[c] Wide range of amplitude and frequency modulation facilities; AM 0 - 99%. DC to 100kHz modulation frequency. FM 20Hz to 100kHz, 99kHz deviation. It is important for centre frequency stability to be preserved on FM.

[d] Level control to give 0 to 139. 9dB in 0. 1dB steps below 2 volts emf from 50Ω.

[e] Should have capability for future extensions e. g. different forms of modulation and provision for a tracking frequency output for measurement system applications as described in the Introduction.

When reviewing the above requirements the need for fast frequency switching indicates that narrow bandwidth phase lock loops cannot be used where large frequency shifts are required. This means that no reliance can be put on these loops to filter any multiplied phase noise from the various standard frequency inputs derived from the frequency standard.

Similarly, the broad range of modulation requirements, particularly FM, suggests a heterodyne approach to allow modulation to be applied at a fixed frequency, thereby achieving a much higher modulation fidelity.

Both of the above points preclude the use of fundamental frequency synthesizers i. e. the type which relies on an oscillator, or oscillator plus signal dividers which actually operate at the output frequency and are locked to some frequency control system. This approach can be useful for CW requirements where switching speed may be slow, e. g. transmitter or receiver driver units.

In order to illustrate the design philosophy of the new 180MHz frequency synthesizer, the UHF Loop and Output circuits will be described in detail.

UHF Loop and Output Circuits of the 180MHz Frequency Synthesizer

The block diagram of this section is shown in Figure 2.

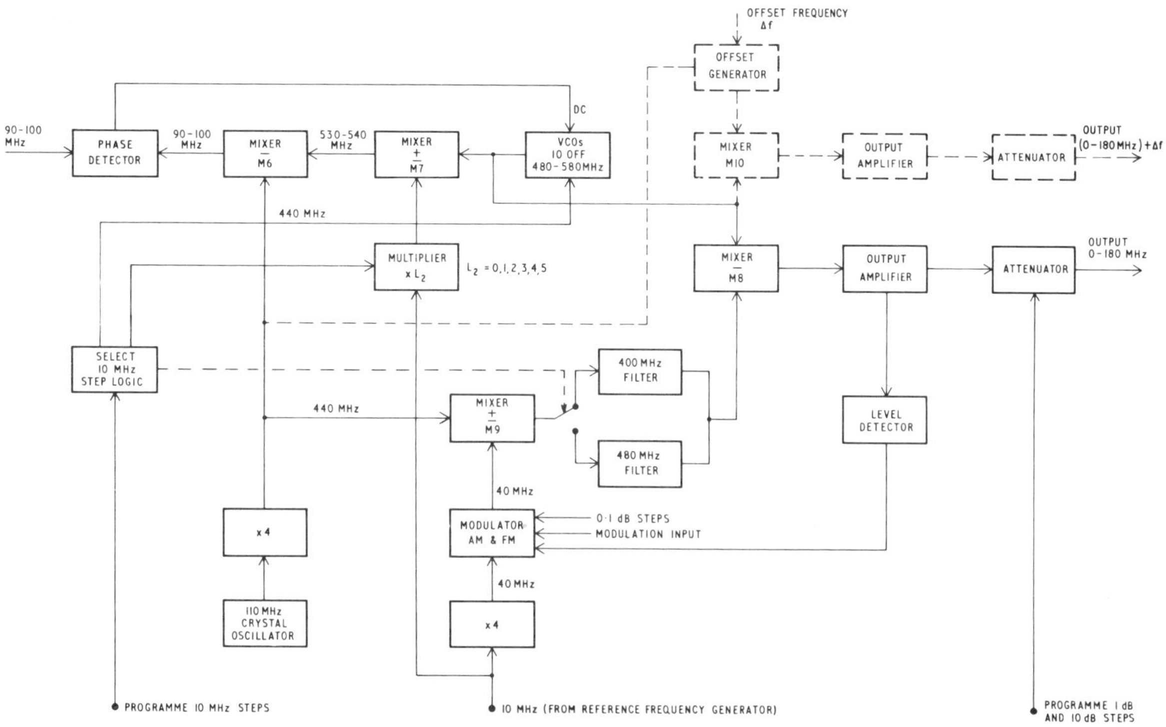

FIGURE 2 4601 - UHF PHASE LOOP AND OUTPUT CIRCUITS

Ten voltage controlled oscillators [VCO] cover the range 480 to 580MHz and are selected, together with multiplier ratio L2, by the 10MHz step programming information. The output of multiplier L2, covering the range 10 to 50MHz steps, is either added to or subtracted from the VCO output in mixer M7 to produce an output frequency in the range 530 to 540MHz. Mixer M6 subtracts a fixed 440MHz from this range of frequencies to produce an output in the range 90 to 100MHz. A phase detector has this signal as one input and the second input is taken from the output of a 90 to 100MHz synthesizer which is variable in 1Hz increments. A phase lock loop is completed by taking the DC output from the phase detector to the control input of the VCO's. The VCO frequencies may therefore be set to 10MHz increments by variation of L2 and 1Hz increments by variation of the 90 to 100MHz input. The 10MHz steps are synchronised to the main synthesizer frequency reference.

It is important to note that in this particular phase lock loop, the highest multiplication is 5[i.e. = 14dB] and therefore the noise of the 10MHz reference is only degraded by this factor when a reference frequency of 50MHz is applied to mixer M7. Because the phase lock loop bandwidth is very broad [≏ 500kHz] the 480 to 580MHz VCO phase noise sidebands are attenuated to the sum of this noise level and a similar contribution from to 90 to 100MHz input.

The 440MHz source, which is produced by multiplying the output of a 110MHz crystal oscillator by four also provides one input for mixer M9. The second input is a 40MHz signal which is derived from the 10MHz signal produced by the frequency reference. Mixer M9 will give both sum and difference frequencies of 480MHz respecively and these frequencies are selected by switched filters. The selection of 400 or 480MHz is made by the 10MHz step control logic.

The output frequency range of 0 to 180MHz is produced by mixing the VCO output frequency with either the 400 or 480MHz fixed signals in mixer M8. An output amplifier followed by a programmable attenuator completes the output path of the synthesizer. The 40MHz signal provides a convenient input for both AM and FM modulation and analogue frequency sweep. The AM facility is put to good use to provide 0.1dB steps in attenuation and also a control input for A.G.C. of the signal output level. A novel circuit allows overall leveling of the output to $\pm$ 0.25dB over the complete frequency range, whilst preserving the facility of DC amplitude modulation.

It should be noted that frequency variations of the 440MHz signal are translated directly onto the VCO frequency as well as the fixed 400 and 480MHz signals. These variations therefore cancel in mixer M8 and do not appear on the synthesizer output. There is therefore no need to synchronise the 110MHz crystal oscillator, from which the 440MHz signal is derived, to the main frequency reference of the synthesizer. For the same reason and because the phase lock loop bandwidth is wide, phase variations due to noise on the 440MHz source also cancel at the output.

Because the earlier stages of the synthesizer cover restricted frequency ranges there is no difficulty in achieving a superior noise performance to that of the UHF Loop and Output circuits. As an example, the theoretical noise limitations of this section can be predicted at 100kHz from the carrier by taking the corresponding phase noise contribution of the 10MHz frequency standard which is -170dB/Hz. The limiting levels are therefore:

[a] Output of multiplier L2 when set for x5 = -170dB/Hz + 14dB = -156dB/Hz.

[b] Output of x 4 multiplier = -170db/Hz + 12dB = -158dB/Hz.

[c] Phase noise contribution from 90 - 100MHz stage = -150dB/Hz.

The combined effect of [a], [b] and [c] is to give a threshold noise limit of approximately -148dB/Hz.

In pratice these levels are limited by amplifier noise figures etc. to give the typical phase noise level of -143dB/Hz.

The design of mixer M8 is particularly critical to preserve a low level of spurious frequency generation. In practice to obtain very good linearity the level of the 480 to 580MHz signal into mixer M8 is amplified to a level of approximately 100mW. This ensures that the level of spurious components generated is considerably lower than 80dB.

Additional Facilities of the 180MHz Frequency Synthesizer

The convenience of providing modulation on a fixed 40MHz subcarrier is particularly useful in offering alternative forms of modulation. The system shown in Figure 2 has already been used for the following forms of modulation:

[a] Single sideband using a complex modulating signal.

[b] Frequency shift keying [F.S.K.].

[c] Amplitude modulation at frequencies up to 300KHz.

[d] Frequency modulation with a modulation frequency up to 500KHz with deviations of 250KHz. The centre frequency being locked to the 10MHz frequency standard.

The most useful feature of the synthesizer however is the capability of providing an additional output signal which tracks the main signal by some defined offset; this being particularly useful for the various two signal measurements given in the Introduction. Figure 2 shows the additional block functions required to provide this facility. A further mixer has as its inputs a signal from the 480 to 580MHz voltage controlled oscillators and a signal from the offset generator. This circuit provides frequencies at 400 or 480MHz but offset in frequency by Δ f. The mixer output frequency is therefore equal to the main synthesizer output but offset in frequency by Δ f. As can be seen this extra facility is very simple and in practice meets the two signal test requirements at a very much lower cost than that for two synthesizers.

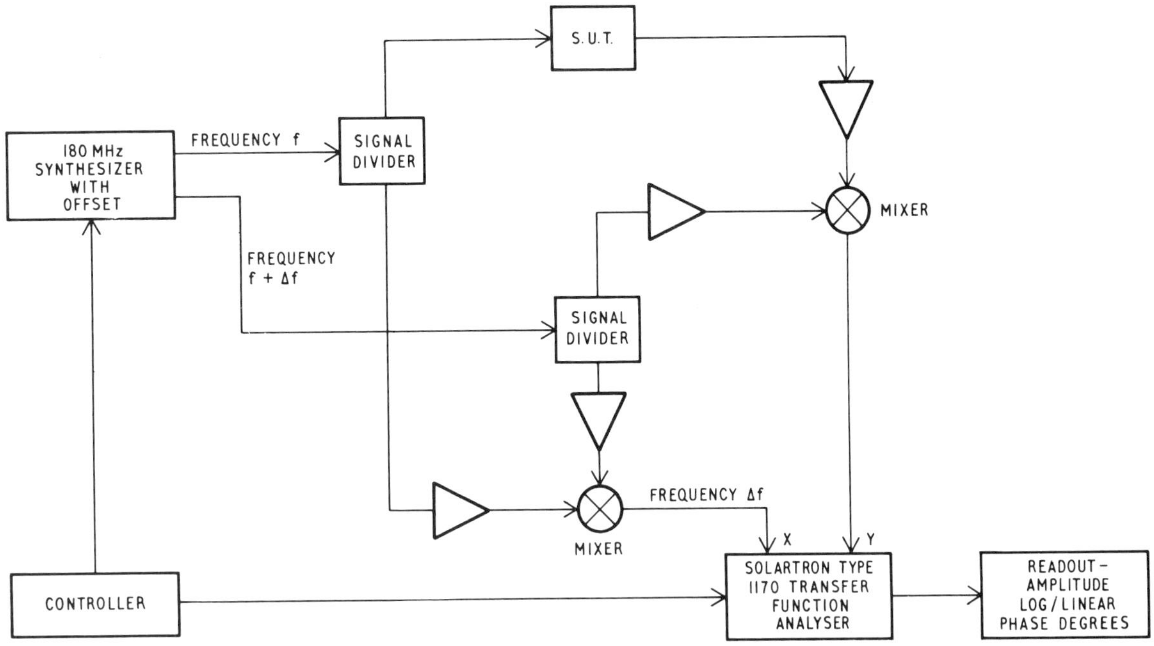

FIGURE 3 EXPERIMENTAL GAIN/PHASE MEASUREMENT SYSTEM

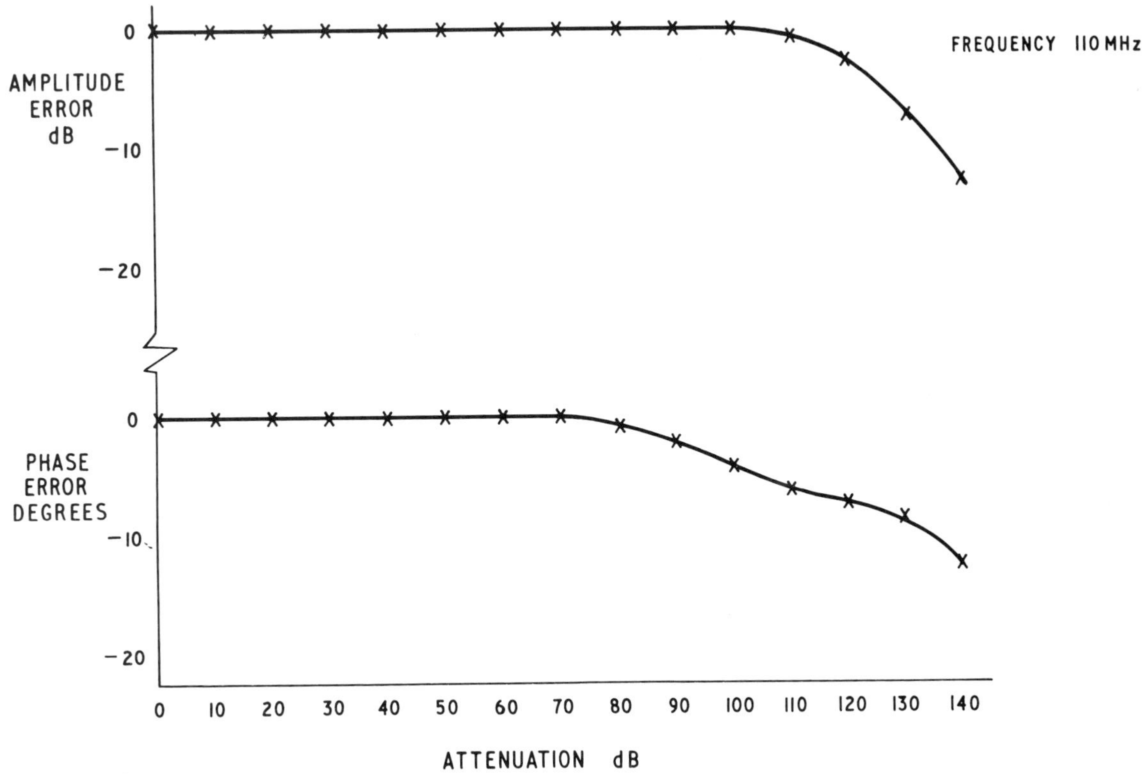

FIGURE 4 PERFORMANCE OF EXPERIMENTAL GAIN/PHASE MEASUREMENT SYSTEM

Experimental Gain and Phase Measurement System

As a further example of the use of a synthesizer with offset capabilities an experimental gain and phase measuring system has been constructed. Figure 3 shows the block diagram of this system; the main synthesizer output is used as the stimulus for the system under test. The offset signal is then used as a reference to heterodyne the input and output signals of the system under test to a fixed low frequency equal to the offset. Since phase and amplitude information due to the system under test are preserved on this low frequency signal all measurements may be conveniently made at this point in the system. For this purpose a commercial Transfer Function Analyser is used. This device uses digital techniques to offer a very wide dynamic range and very high sensitivity.

The complete experimental system has a sensitivity of better than -120dBm and Figure 4 shows measurements made on a precision attenuator over a range of 140dB. It is particularly significant to note that quite consistent phase measurements have been made over the complete range of attenuation. This is thought to be unique in an instrument of this type. The errors shown in Figure 4 are mainly due to signal breakthrough and work is in hand to reduce this contribution.

Acknowledgements

The author wishes to thank the R. & D. Director of Solartron - Schlumberger Ltd. for permission to present this paper and to the members of the R. F. Instrument Group for their efforts in developing the instruments described.

DIGITAL WAVEFORM SYNTHESIS

E. Jenks and E. T. Powner

Introduction

In recent years there has been a growing interest in the production of waveforms of a given shape and many methods have been utilised to generate such functions. Although the method of digital synthesis is fast becoming the most popular, the majority of digital synthesis methods used to date have the disadvantage that, whilst it is a relatively simple matter to generate a given waveform, any change in this waveform requires the replacement of a large proportion of the whole system. This means that the production of a general-purpose waveform synthesiser is relatively expensive.

The system described here is capable of generating waveforms of almost any waveshape and, although the initial cost is higher than in many comparable single-function systems, changing the output waveform is a relatively simple and cheap operation. Such a change, which can be made programmable, requires only a minimum of alterations to the existing circuitry.

The Principle of Digital Systems

The principles of digital synthesis are well known and hence only a brief resume will be given here. Consider, for the sake of illustration, a sine wave; this can be divided into a number of sections each of variable amplitude (see Fig.1). Each section of the waveform can thus be represented by a constant width, variable amplitude pulse.

The problem now arises of how many pulses should be used to represent one cycle of the waveform and it has been found in practice that, for the simple example above, at least 12 pulses per cycle are necessary to produce a waveform with an acceptable degree of distortion (Powner, Green and Taylor (1)). Hence any digital waveform synthesiser should produce a fixed number of pulses per cycle (dependent on the accuracy required), each of which is of variable amplitude such that the pulses synthesise into the desired waveshape.

The Look-up table

Such a system may be very simply and accurately produced by using a reading method in the form of a look-up table as shown in Fig.2. Here the amplitude of each pulse comprising the wave is stored in a Read-Only-Memory (R.O.M.). The waveform is produced by reading these amplitudes from the memory in a cyclic manner and gating the pulses so formed via a digital-to-analogue converter and a sample and hold to give the required analogue waveshape. Such a system, treated in more detail by Evans (2), is capable of giving a high degree of accuracy dependent on the number of bits used and the number of storage locations in the R.O.M. Waveshape

Mr. E. Jenks and Dr. E.T. Powner are both with U.M.I.S.T.

changing, however, is inconvenient as it involves either a complete new R.O.M. (in the case of non-reprogrammable types) or the modification of those which may be to some extent reprogrammable. The first alternative is expensive and the second is an operation which may be tedious and involve a time penalty, as well as implying the loss of the original data. Hence if a look-up table system is to be used for a waveform synthesiser of more than one wave output either a multi-R.O.M. system is required or some alternative method for storing the look-up values becomes necessary.

One way of achieving the latter is by use of a Random Access Memory (R.A.M.), such a memory allowing new data to be input at will. This overcomes most of the problems associated with R.O.M.'s but the problem of loss of original data when reading in a new set of conditions, still remains. The authors overcome this problem by use of a variable-code generator which generates a digital code indicative of the amplitude of pulses required to form many of the synthesised waveforms, the code being stored in the R.A.M.

Variable Code Generation

The system briefly outlined above requires circuitry capable of sequencing the input to the R.A.M. such that when read out cyclically, pulses appear as a digital representation of the required waveform . This could, of course, be done by manual transferring of the pulses each time a change of waveshape is required but such a system has little advantage over the reprogramming of a R.O.M. Hence in order to generate the necessary pulses automatically a cyclic sequence generator such as those based on shift registers is required.

The use of shift register sequences to generate known distributions of states (described statistically) has been covered in literature (e.g. Heath and Gribble (3)), and it is well known that if feedback is applied to a n-stage shift register it is possible to generate a cyclic sequence of maximum length 2^n states.

Such systems, however, generate each state only once during each complete cycle and thus if a distribution of states representing the output waveform is to be generated the occurrence of each state will need to be heavily weighted by the encoding logic.

To avoid such weighting systems in the encoding logic a sequence which provides the multi-occurrence of the same state in one cycle is necessary. Such a sequence may be generated by use of a 'window' method in which the output from only a section of the shift register is used to form the necessary pulses. Again, however, there is a tendency to a uniformity of distribution over a complete cycle.

Ideally what is required is a multi-occurrence system in which the distribution of states over a complete cycle is non-uniform so that the encoding logic need consist of inhibit lines to suppress certain states and the logic necessary to input the remaining states into the R.A.M. in such a way that, when output cyclically, they appear in the correct sequence to form the required waveshape.

The Corrupted Code Generator

To produce a device which has multi-occurrence of states without uniformity of distribution the authors have been investigating the use of a system they describe as a 'corrupted code generator'. Here a feedback shift register is used to produce the required code but each stage of the register does not necessarily receive a clock pulse at each clock-time. The arrival of clock pulses to each stage of the register, known as the primary register, is controlled by the state of a second shift register known as the control register. A diagram of a typical system with a 5 stage primary register being clocked from a 4 stage control register is given in Fig.3.

A configuration such as that shown has a maximum possible cycle-length of $2^5 \times 2^4$ (= 496) (i.e. the same as a full 9 stage register) but the distribution of states over a complete cycle may be non-uniform, giving the kind of multi-occurrence state distributions deemed necessary for the variable-code generator.

The authors have investigated the use of many different systems of feedback in both the primary and central registers by use of a software simulation of the system. The results obtained from the simulation show that a simple change in the control register or the rearrangement of the connections between the two registers is all that is often required to produce sequences with different state-occurrence distributions. By careful selection of the encoding logic used it has been found possible to generate a number of different, useful waveforms by means of the simple alterations to the basic registers outlined above. In a practical system such changes may easily be made by use of switches or interchangeable plug-in boards.

For a very simple example of the changes possible consider the two configurations given in Fig.4. The only difference between these two systems is the feedback on the control register but they produce widely different state-occurrence histograms as shown in Fig.5. Using identical encoding logic on each system such that only states of decimal decode 28,23,25,29, 24,10 and 20 are input into the R.A.M. (cyclically in that order) it will be found that system (a) gives an output of a sawtooth wave whilst system (b) gives a triangular wave (see Fig. 6).

If the function required is not available by changes such as those used above it is possible to change the primary register with little difficulty and, if a major function change is required, the whole encoding logic may be changed to suit.

The frequency of the output waveform is determined by the frequency at which the R.A.M. is accessed (being limited by the minimum access time of the memory) whilst the amplitude of the output is controlled by connecting a variable gain amplifier to the output of the sample-and-hold. It will be noticed that the input to the R.A.M. need not be rewritten for each output cycle but need only be input once by use of (say) an 'initialise' switch which applies the clock for long enough to complete one cycle of the corrupted code generator.

The distortion in the output waveform is defined by the number of pulses in each corrupted code cycle, more pulses causing less distortion and vice versa. The authors consider that, typically, 100 pulses would be required as a minimum to give an acceptable degree of distortion and it has been found necessary to use a corrupted code generator with a 5 stage primary and 4 stage control register to achieve sequences of this length.

Conclusion

The system described above is highly versatile with the minimum of changes being necessary to obtain a great variety of different functions. A full block diagram of a digital frequency synthesiser is given in Fig.7.

The authors are continuing to use their software simulation of the corrupted code generator to investigate various other combinations of primary and control registers and it is hoped to produce some useful sequences using higher order registers (e.g. 6 stage primary, 5 stage control). Such investigations must, however, be considered as a long term project.

References

1. Powner, E.T., Green, D.H., Taylor, G.J.: Digital Waveform Synthesis, Electronic Engineering, August 1969.

2. Evans, W.A.: Waveform Synthesis, IEE School on Electronic Instrumentation, April 1973, p.12/1-12/13.

3. Heath, F.G., Gribble, M.W.: Chain codes and their Electronic Applications, Proceedings IEE, Vol.108C, March 1961, p.50-58.

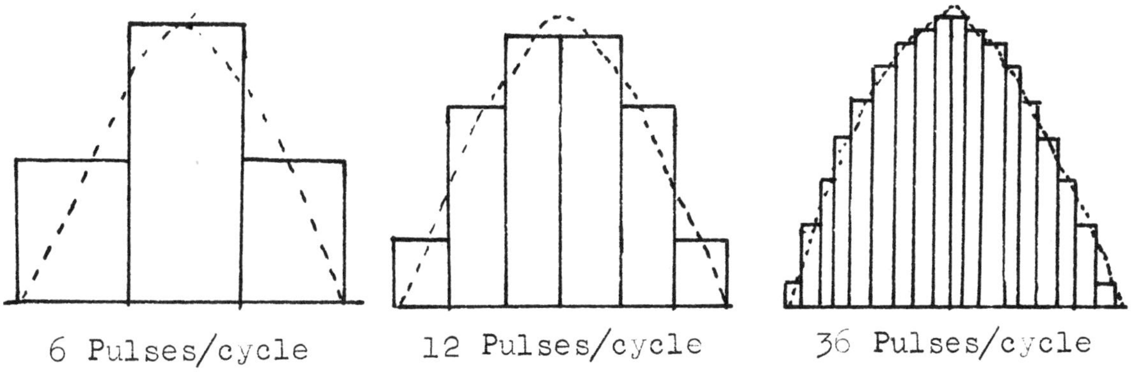

Fig.1. Comparison of Varying Pulses/cycle

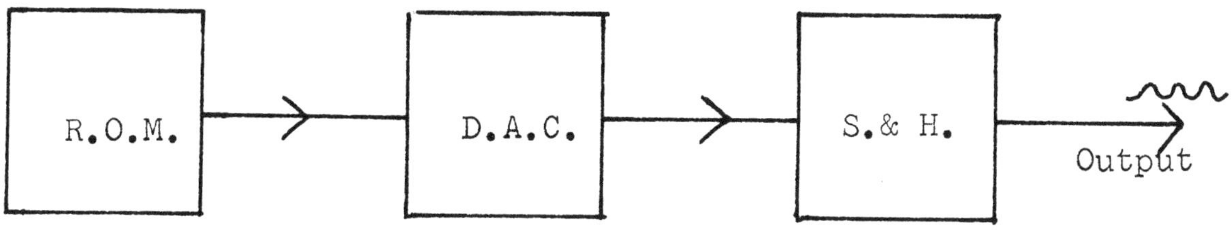

Fig.2. Look-up Table Synthesis

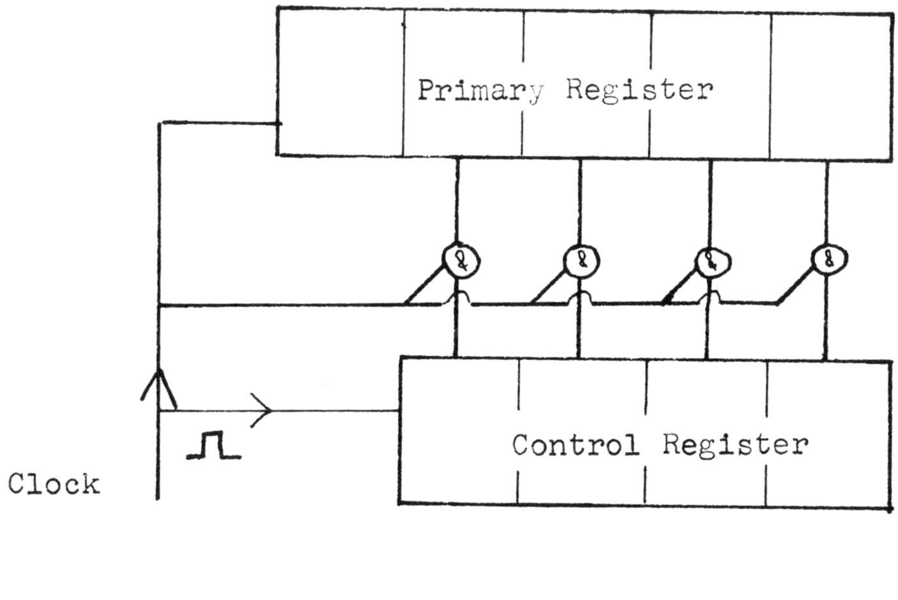

Fig.3. Corrupted Code Generator

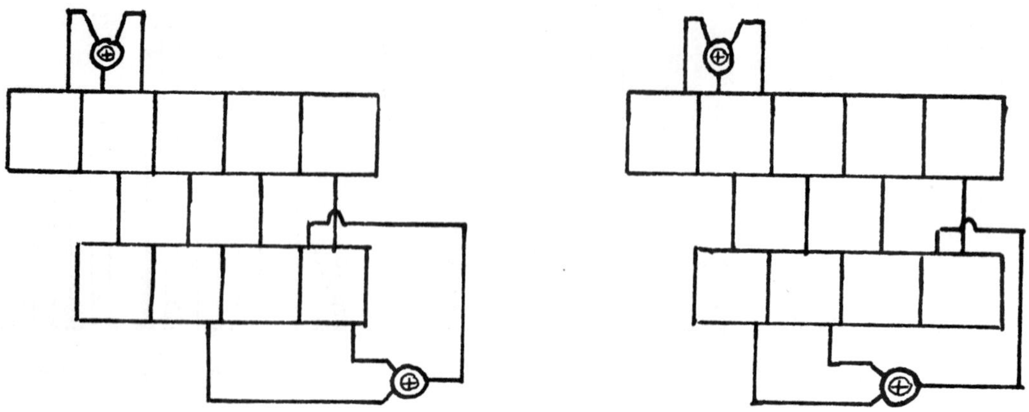

(a) PRIMARY B=A⊕B

CONTROL D=B⊕D

(a) PRIMARY B=A⊕B

CONTROL D=A⊕B

Fig.4. Configurations in example

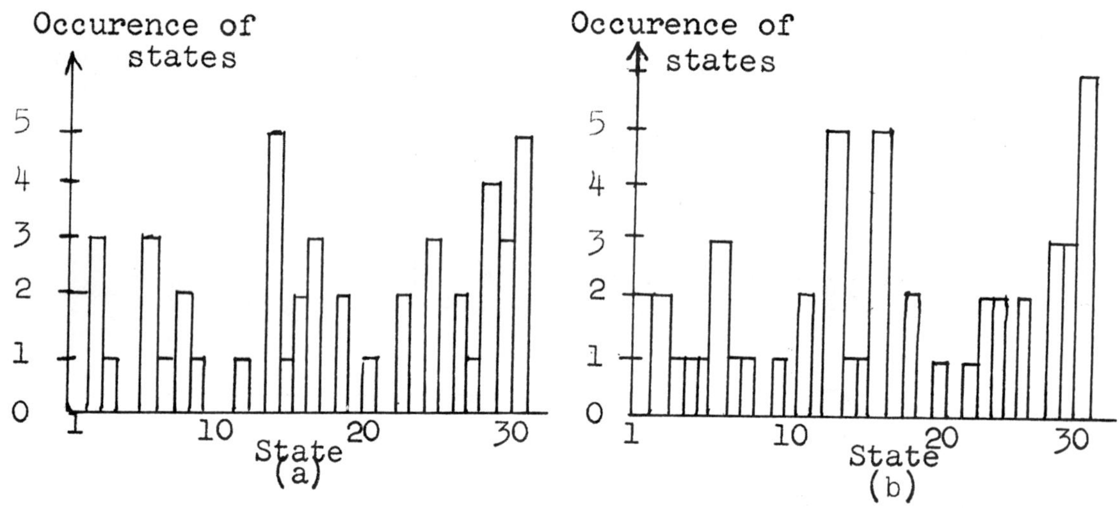

Fig.5. State Occurence Distributions

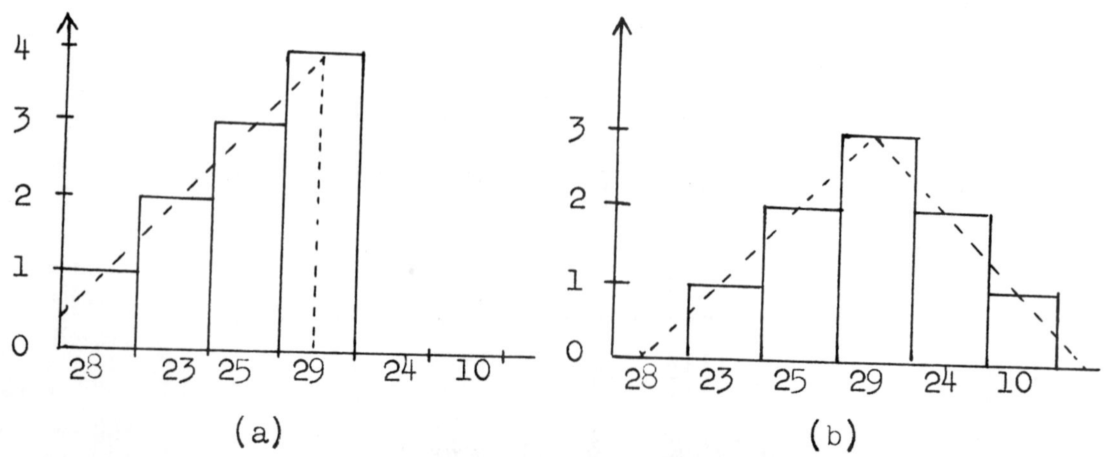

Fig.6. Output Waveforms

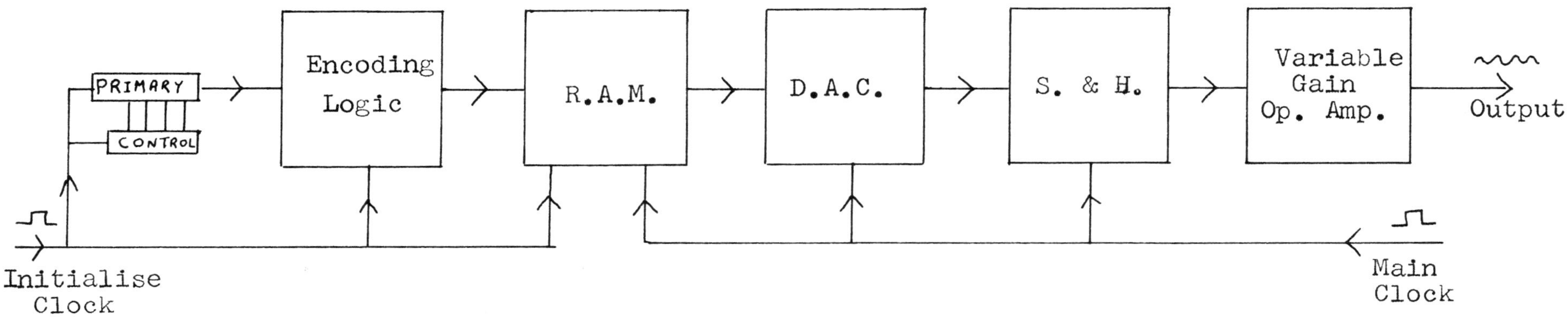

Fig.7. Complete System

PROGRAMMABLE PRECISION SIGNAL GENERATORS

M.B. Ashdown

1. Introduction

Testing of modern communications equipment requires signal generators having accurately controlled frequencies coupled with very good spectral purity. In addition the increasing use of automatic test methods require these signal generators to be capable of computer control of all functions. This inevitably means the use of a synthesised signal source for the generator. Unfortunately conventional techniques of frequency synthesis produces unwanted signals at levels too high for the performance testing of modern communication receivers. It will be shown later that spurious signals must be at least 100dB below the wanted carrier, which is at least 20dB better than is attainable by normal synthesis techniques.

Although many measurements may be made using a single signal source, the most searching measurements, such as cross modulation, blocking, reciprocal mixing and intermodulation, require two independent tones, each capable of covering the full frequency range. This could be accomplished by using two separate signal generators or a single composite generator, but it is important that the combining of the two tones does not in itself produce unwanted spurious signals.

2. Limitations of Conventional Synthesisers

A block diagram of a hypothetical synthesised signal generator using conventional techniques is shown in Fig. 1. This generator covers the range 0 - 150MHz, and from this it can be seen that the output frequency is derived by mixing a 400MHz - 550MHz synthesised signal with a fixed 400MHz reference derived from the frequency standard. A simple calculation will show a serious 5th order intermodulation product centred on 133.1/3MHz

$$
\begin{aligned}
fo &= 533.1/3 - 400 \\
&= 133.1/3 \\
fs &= 3 \times 400 - 2 \times 533.1/3 \\
&= 1200 \quad - 1066.2/3 \\
&= 133.1/3\text{MHz}
\end{aligned}
$$

M.B. Ashdown is Chief Engineer (R.F. Measurement) of Racal Instruments Limited.

Figure 1. A Typical Synthesised Signal Generator.

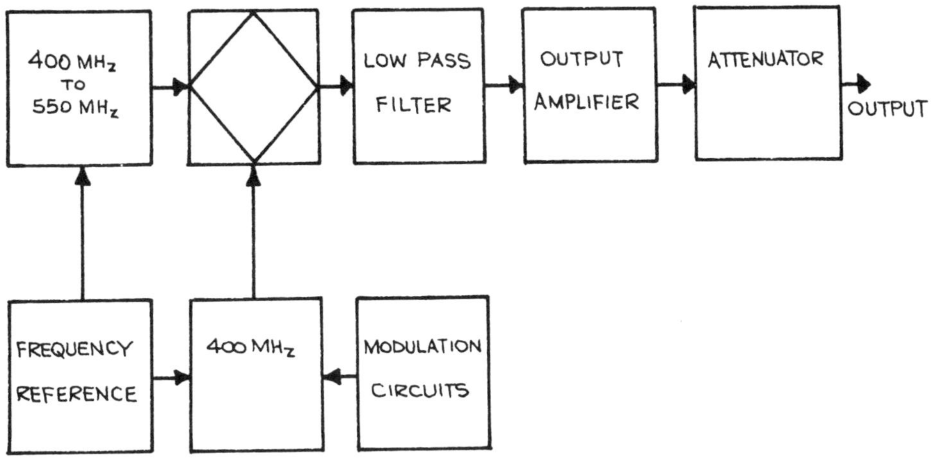

Figure 2. A Digital Synthesised Signal Generator.

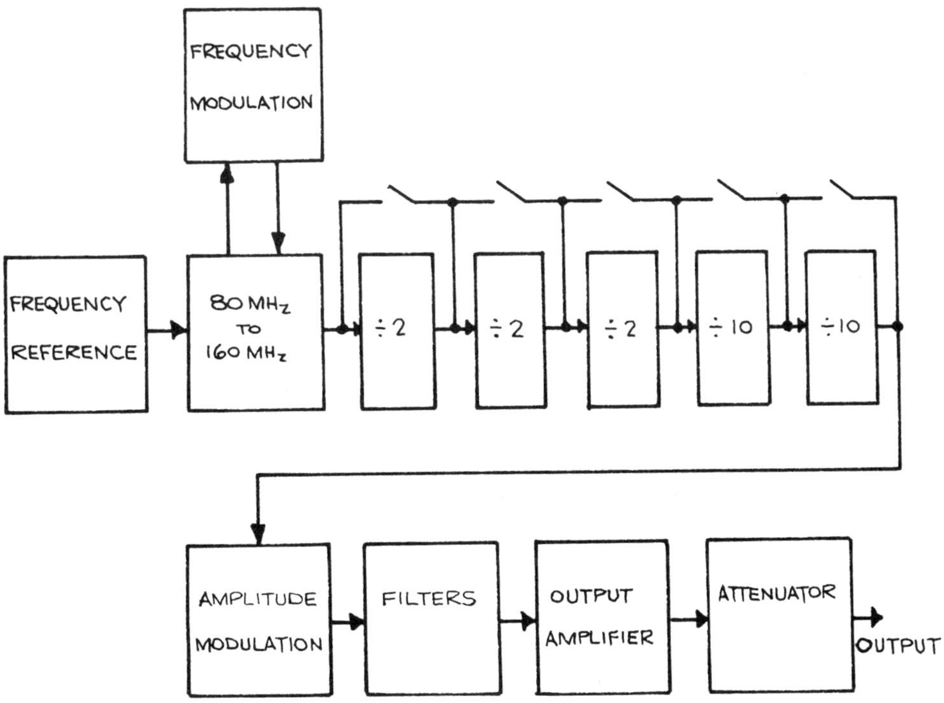

Figure 3. A Typical Digital Phase Locked Loop.

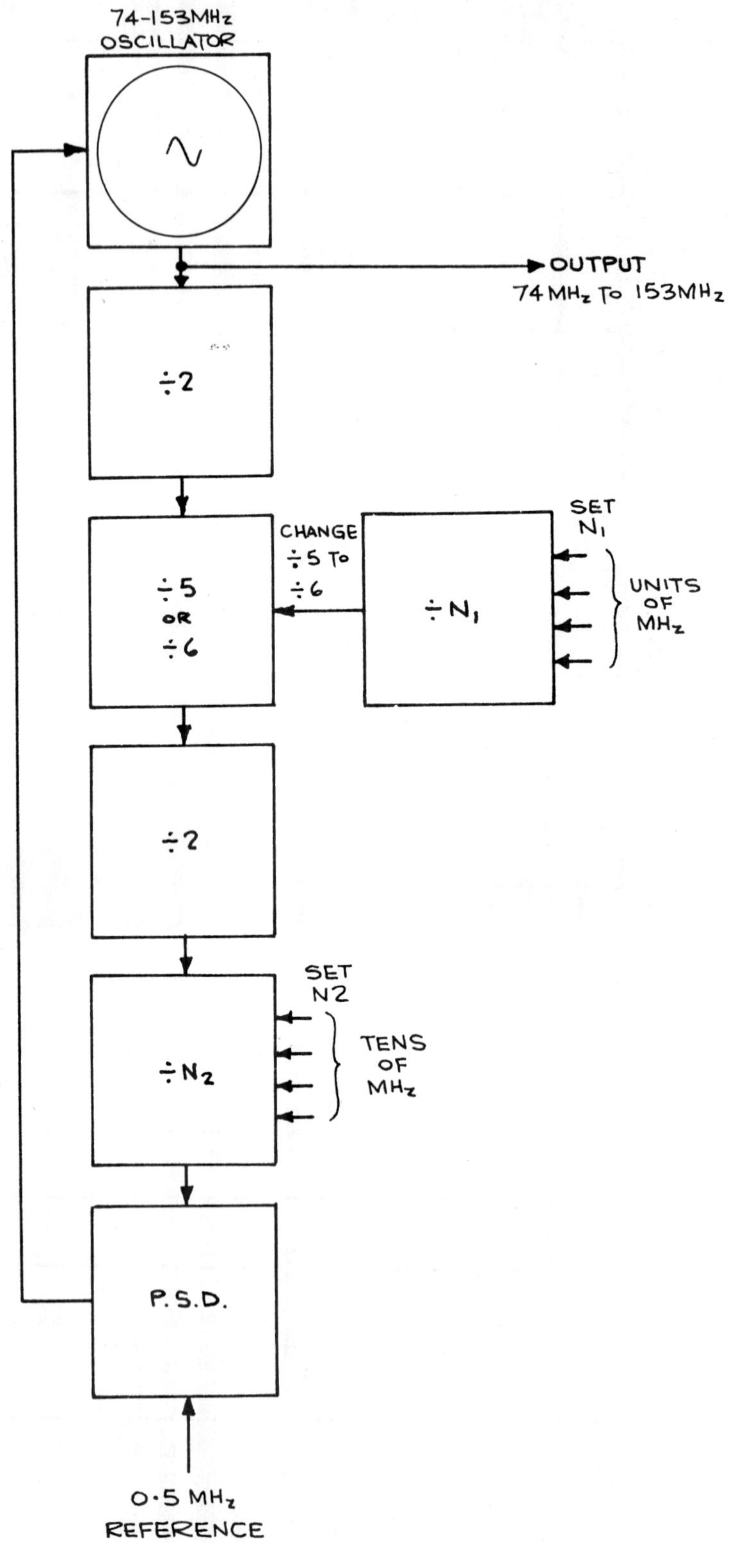

This component can appear anywhere in the output range, for example, if the generator was tuned to 150MHz this spurious component would be at 100MHz. This component would not be troublesome if its level could be controlled to be less than 100dB below the carrier, but even using the best mixers available today, manufacturers do not claim spurious performance of better than -80dB.

Owing to this serious limitation on the overall performance of the signal generator, an alternative solution is clearly called for.

3. Digital Frequency Synthesiser

3.1 Use of Output Divider

The alternative technique which has been adopted makes use of an octave tuned synthesiser covering the range 80MHz - 160MHz, followed by a series of binary and decimal digital dividers (see Fig. 2). As can be seen by the use of only three binary and two decimal dividers, an output range of 0.1MHz - 160MHz can be produced. The technique, although new for synthesisers, has been adopted on a number of modern wide range manual signal generators. In addition to eliminating the output mixer and hence the wide band spurious problem, a further important advantage of the technique is that close-in noise is reduced by successive division.

3.2 The Kay Multiplier

It is apparent that whatever output frequency is demanded the synthesiser can only cover the range 80MHz - 160MHz, and all lower frequencies require the use of the output dividers. In order to relate synthesiser frequency, division ratio and commanded frequency, a unique logic system called the Kay Multiplier (Ref. 1) has been developed. This system accepts the commanded frequency, selects the required output division ratio and multiplies the BCD representation of the commanded frequency by the same ratio in order to set the synthesiser.

3.3 Digital Phase Locked Loops

The synthesiser is composed of a number of phase locked loops, all digital in concept. A typical example of such a loop is shown in Fig. 3, and as can be seen the output covers the range 74 - 153MHz in 1MHz steps. At lower frequencies a loop such as this presents no problems, but as the frequency is increased the techniques for the programmable divider become more complex. The dividers shown use a variation of the swallow counter to produce division ratios that are effectively irrational numbers. The manner in which this divider operates is best illustrated by an example. Thus if the desired output frequency is 125MHz the dividers are set as follows:-

Set N_1 to full length 24 and change point 5

Set N_2 to 12

Now the $\div 5/\div 6$ is arranged to divide by 6 up to the change point determined by N_1, and thence to divide by 5 up to full length of N_1.

Thus output pulses from $\div 5/6 = 24$ for full cycle

input pulses to $\div 5/6 = 5 \times 6 + 19 \times 5$

$= 125$

$\therefore$ Effective division ratio $= \frac{125}{24}$

$\therefore$ Total division ratio $= \frac{125}{24} .4 .12$

$= 250$

Thus output frequency $= 250 \times 0.5\text{MHz}$

$= 125\text{MHz}$

4. Two Tone Measurements

The four most important two tone measurements required on H.F. communications receivers are:-

(i) Cross-Modulation

(ii) Blocking

(iii) Reciprocal mixing

(iv) Intermodulation

These four measurements will now be dealt with in turn. Consideration will be given to the method of measurement, the expected performance from a modern receiver and the performance requirements of the signal generator.

4.1 Cross-Modulation

Cross-Modulation occurs when the modulation signal of an interfering carrier modulates the carrier of the wanted signal.

It should be noted that cross-modulation can only be caused by third or higher order curvature of the active elements in the receiver.

4.1.1 Method of Measurement

Measurements may be made with the receiver in the a.m. or SSB mode. In the a.m. mode the wanted signal is set to a carrier level of say 1mV and modulated to 30% by a 1kHz signal. The 1kHz demodulated tone in the receiver is set to a convenient level, e.g. 1mW.

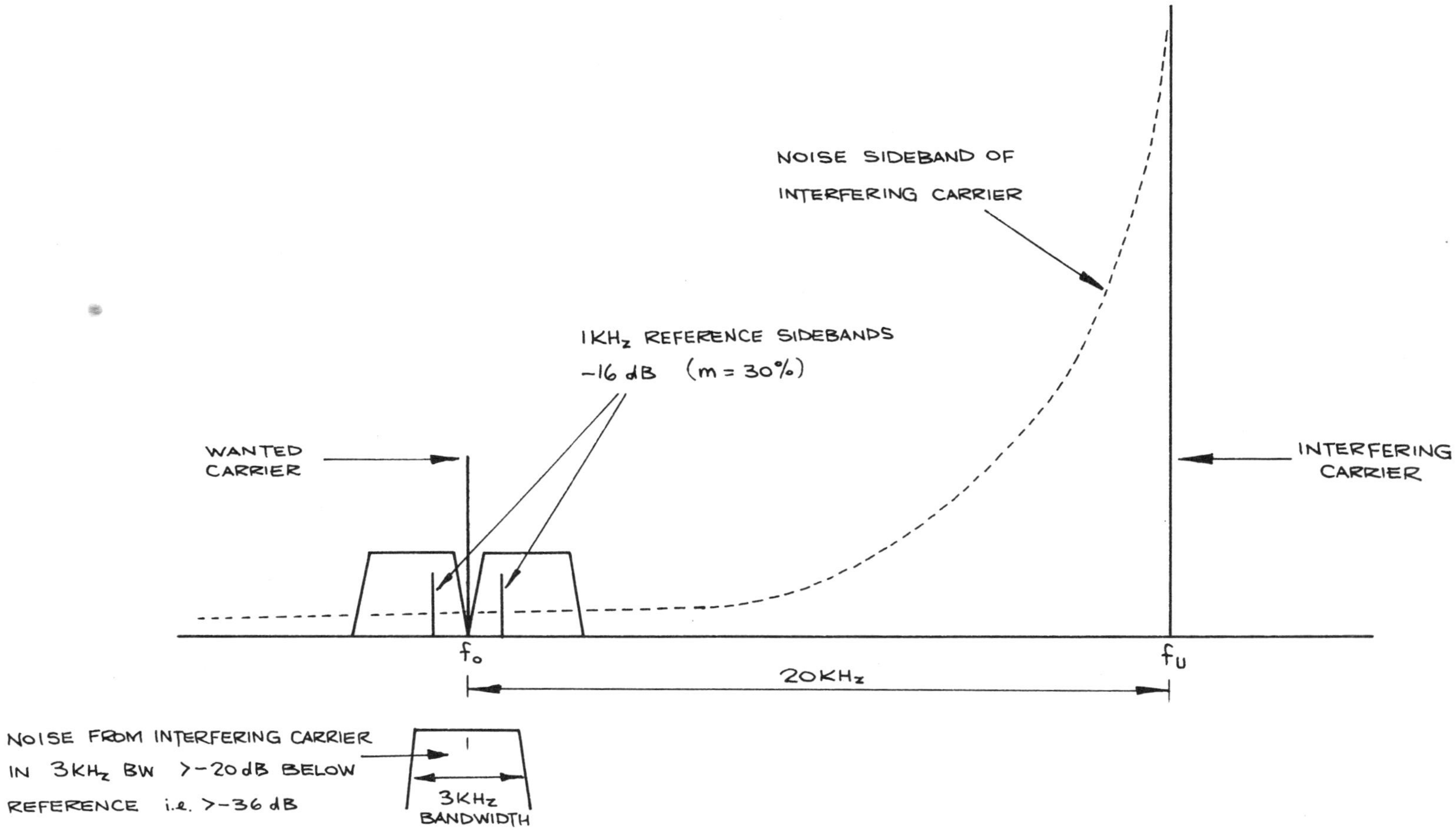

Figure 4. Effect of Signal Generator Noise on Cross Modulation Measurement.

The modulation is then switched off and an interfering signal, out of band of the I.F. but inside the selectivity band of the R.F.circuits, is switched on (the frequency is typically 20kHz off tune). The interfering signal is modulated to 30% by a 1kHz signal and the carrier amplitude is increased until the demodulated 1kHz tone is -20dBm (this corresponds to a cross-modulation of 10%, i.e. a 30% modulated interfering signal produces 3% modulation of the wanted carrier).

For a high quality receiver the level of the interfering carrier should be at least 0.5V. It should be noted that the cross-modulation measurement can easily be masked by blocking (see 4.2 below) or reciprocal mixing (4.3 below), if the receiver suffers badly in these respects.

4.1.2 Generator Performance

To make the cross-modulation measurement, the signal/noise ratio of the generator, 20kHz from the carrier, must be very good. This is shown in Fig. 4. The sideband level required to produce the equivalent noise of a 3% modulated signal on a 1mV carrier is -36dBm on 1mV. The ratio of wanted to unwanted carrier levels is 500mV to 1mV which, expressed in decibels, is -54dB. Therefore, the S/N ratio of the 500mV signal expressed in a 3kHz bandwidth, 20kHz from the carrier, must be at least 90dB (36dB + 54dB). The noise performance of the signal generator should be 10dB better than this. The signal generator should, therefore, have a S/N ratio of 100dB measured in a 3kHz bandwidth, 20kHz from the carrier (equivalent to 135dB in a 1Hz bandwidth) and corresponding spurious levels at least 100dB below the carrier.

4.2 Blocking

Blocking occurs when an interfering carrier reduces the sensitivity of the receiver to the wanted signal.

Blocking is normally caused by a high level interfering carrier affecting the linearity of the first or second mixer. It can also be caused by the interfering carrier breaking through the I.F. filter and affecting the agc voltage.

4.2.1 Method of Measurement

The measurement may be made on a.m. or SSB. In the a.m. mode the wanted signal is modulated with a 1kHz tone. In the SSB mode the wanted signal is offset to give a 1kHz audio tone. The level of the wanted carrier is set to about 1mV. The interfering signal is set typically 20kHz from the wanted signal and its level is increased until the level of the 1kHz audio tone falls by 3dB.

4.2.2 Generator Performance

For this measurement the generator should be capable of producing an output of 1V e.m.f. The S/N ratio of the generator is not so critical as that required for the cross-modulation measurement.

4.3 Reciprocal Mixing

Reciprocal mixing occurs when the noise of the local oscillator is transferred onto the wanted carrier by the presence of a high level unwanted carrier.

4.3.1 Method of Measurement

Measurements are normally made with the receiver tuned to SSB. The wanted signal is set to a low level (e.g. 10μV), with the agc of the receiver set to manual. The audio tone of (say) 1kHz is adjusted to give a 1mW output. The unwanted tone is offset from the nominal carrier frequency by 20kHz and its level is increased until the S/N ratio of the audio tone is degraded to 20dB. For a good receiver the level of the unwanted tone should be +70dB on the wanted.

4.3.2 Generator Performance

The generator should have a S/N ratio of at least 10dB better than the performance being measured, i.e. 100dB in a 3kHz bandwidth (135dB in a 1Hz bandwidth), 20kHz removed from the carrier. Similarly, any spurious frequencies should be 100dB below the level of the unwanted carrier.

This particular measurement demands a very high performance from the signal generator. Synthesised signal generators are notorious for their poor S/N ratio and high levels of spurious frequencies. Until quite recently the best available spurious performance was -80dB, which is clearly not adequate for the reciprocal mixing test.

4.4 Intermodulation

Intermodulation occurs when two or more signals (in band or out of band), or harmonics of them generated within the receiver, produce an unwanted modulation tone.

The worst intermodulation components are normally the second and third order terms.

The third order terms occur from the sum and difference of the second harmonics and fundamentals of the unwanted carrier:-

e.g. $f_{int} = 2f_1 \pm f_2$, and $2f_2 \pm f_1$

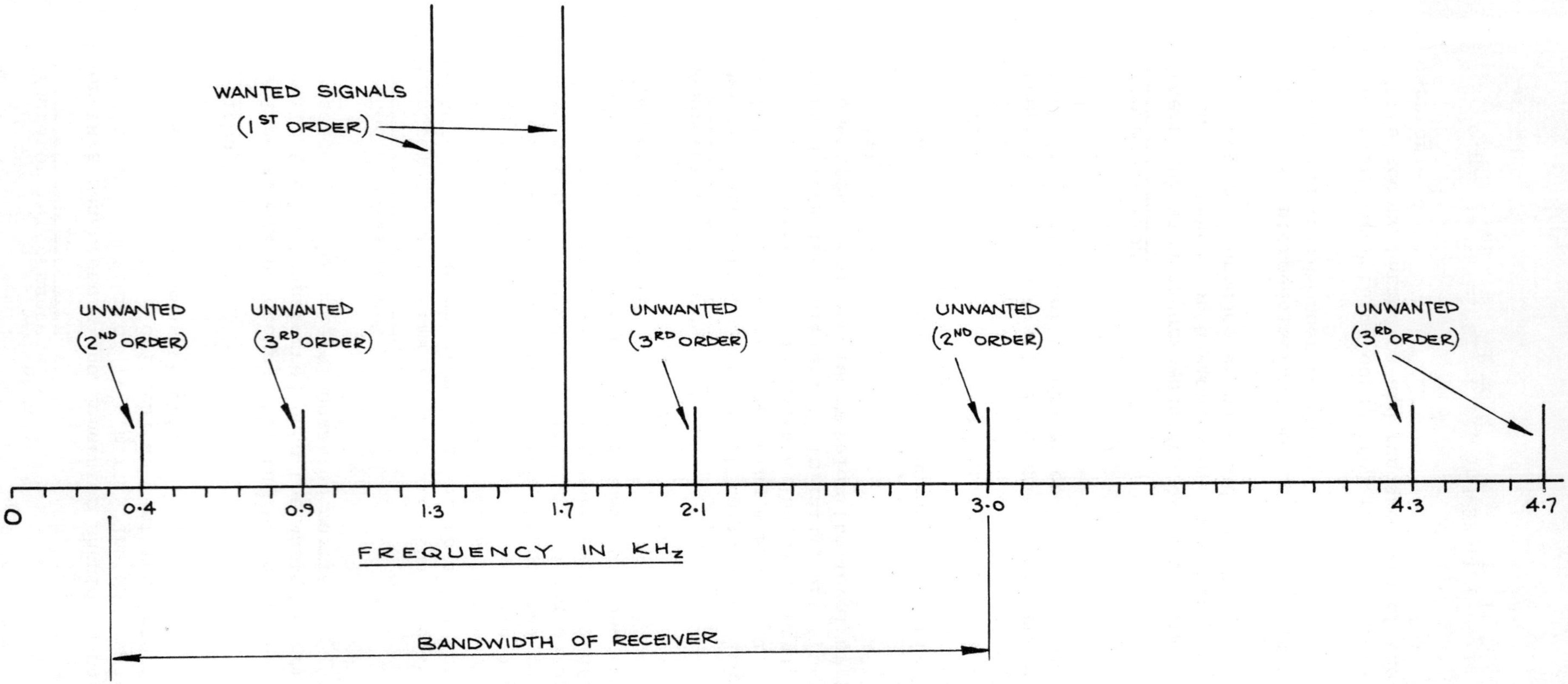

Figure 5. 2nd and 3rd Order Intermodulation Components.

The I.P.s can be expressed in a general form for even and odd order terms respectively:-

For even order terms : $F_{int} = N\Delta$ and Nf_o

For odd order terms : $F_{int} = Nf_o \pm \Delta$ and $f_o \pm N\Delta$

where $\Delta = \frac{f_2 - f_1}{2}$

$f_o = \frac{f_1 + f_2}{2}$

and N = Nth order product.

4.4.1 Method of Measurement (Out of Band)

The receiver is set to USB with a 1μV wanted signal, giving a 1kHz tone at an output level of 1mW. The wanted tone is then switched off.

For the measurement of 2nd order I.P.s tone A is set to f_c + 20kHz, and tone B is set to 2 f_c + 21kHz (where f_c is the nominal carrier frequency).

The I.P. then occurs at:-

$$(2f_c + 21\text{kHz}) - (f_c + 20\text{kHz}) \doteq 1\text{kHz}.$$

The level of the two R.F. tones are increased until the 1kHz I.P. has a level of 1mW.

For the measurement of 3rd order I.P.s tone A is set to f_c + 20kHz and tone B is set to f_c + 30kHz.

The I.P. then occurs at:-

$$2(f_c + 20\text{kHz}) - (f_c + 39\text{kHz}) = 1\text{kHz}.$$

The level of the two R.F. tones are increased, as from the 2nd order measurement, until the 1kHz I.P. has a level of 1mW.

For a high quality communications receiver the levels of the tones should be + 90dB with respect to 1μV.

4.4.2 Method of Measurement (In Band)

The receiver is set to USB. Both R.F. tones are set in-band at frequencies of f_c + 1.3kHz and f_c + 1.7kHz.

The levels of the R.F. tones are increased until the 2nd and 3rd order I.P.s are -50dB with respect to the 1.3kHz andl.7kHz wanted tones (see Fig. 4).

It should be noted that the I.P.s will occur at the following frequencies:-

2nd order terms : 400Hz and 3kHz

3rd order terms : 900Hz, 2.1kHz, 4.3kHz and 4.7kHz (the latter two terms will be out of band).

4.4.3 Generator Performance

For the out-of-band measurement the level of the R.F. tones for a good receiver would have to be +90dB above 1μV.

Therefore the wide band spurious performance of the signal generator (for components > 20kHz from the carrier) should be -100dB, and the S/N ratio measured 20kHz from the carrier should be 135dB/Hz.

For the in-band measurement the spurious levels of the signal generator should be -60dB and the S/N ratio, measured 1kHz from the carrier, should be 95dB/Hz.

5. Reference

5.1 'A 160 MHz Synthesised Signal Generator for the Automatic Testing of Radio Receivers'.
by K.R. Thrower CEng. MIERE
Presented at the joint Conference on Radio Receivers at Swansea University in July 1972 (IERE Conference Proceedings No.24).

FREQUENCY SCALING, BY DIGITAL MEANS

J.D. Martin

1. Introduction

Pulse-frequency (or pulse-rate) is often used to represent a signal variable in instrumentation applications, because many measuring transducers naturally produce this kind of ourput. A signal in pulse-frequency form can readily be manipulated by wholly digital circuits, thus capitalising on the ease of construction, stability and ruggedness of digital circuits. Scaling, that is multiplication of a pulse-frequency by a constant, is often at the heart of more elaborate systems,[1, 2] and although several ingenious arrangements have been devised for this duty, the limitations inherent in this operation are not always obvious. A general discussion of the problem is presented here, together with a brief survey of available practical techniques.

Frequency scaling is the essence also of frequency synthesizer design and it would be attractive to produce a wholly digital system which will scale the frequency of a stable crystal oscillator to give a variable output frequency. Although such applications will be mentioned here, a high purity synthesizer requires the use of analog circuit techniques to interpolate smoothly on the frequency scale. Such arrangements, based on phase-locked-loops offer attractive and reliable solutions, but are specifically excluded from the present discussion.

2. The basic scaling operation

Digital systems process a pulse-rate signal by manipulating period; which is in contrast to analog systems which by controlling the frequency of an output oscillator, manipulate frequency directly. Consequently in a digital system, it is impossible to obtain a frequency proportional to the controlling parameter.

Consider a scaling arrangement where a constant frequency source is available of frequency f_i, having a period τ. An output frequency f_o is required, having a frequency $f_o = a.f_i$. $(a \leq 1)$

Now, for practical values of a, this may be expressed as a ratio of integers, thus:

$$a = M/N \qquad (1)$$

The output period is given by:

$$\frac{1}{f_o} = \tau(Q + \frac{R}{M}) \qquad (2)$$

where Q is the quotient of N/M, and R is the remainder: $(N \geq M)$. When the remainder R is zero, then exact scaling is possible, but is restricted to output frequencies of the form f_i/Q, where Q is an integer. This is the simple case of straight division of the input frequency by a counter. However, for all other cases, exact scaling is impossible, and a choice must be made between an irregular pulse period with the average frequency being accurately proportional to the controlling parameter <u>a</u>, and a regular pulse period whose frequency is the nearest integer division of the input frequency.

J.D. Martin is with the School of Electrical Engineering, University of Bath.

2.1 Irregular pulse period

Given the proposition in equation 2, M output pulses can always be produced in a period defined by N input pulses, giving an exact average frequency of $\frac{M}{N\tau}$. Output pulses can only be generated at times which are multiples of τ, since no other timing information exists between these instants. The correct average output frequency can be generated by allowing R output periods of $(Q + 1)\tau$ and $(M - R)$ periods of $Q\tau$ for every N periods of the input frequency. A little simple algebra will show that this can always be achieved no matter what the relative values of M and N.

As well as ensuring the average frequency, it will be desirable to keep the instantaneous frequency as close to the ideal as possible, and this requires that instantaneous period error be minimised. Figure 1 shows a simple algorithm which carries out this operation; RR is a register which holds a cumulative remainder. It is not difficult to realise this type of algorithm in either special digital hardware, or in software.

Consider now the instantaneous fluctuations, which give rise to errors in instantaneous-period and instantaneous-frequency. The actual periods are $Q\tau$ and $(Q + 1)\tau$, while the desired period is $\frac{N\tau}{M}$. The relative period errors are then given by:

$$\epsilon_p(-) = \frac{-R}{N} \qquad \epsilon_p(+) = \frac{M}{N} - \frac{R}{N} \tag{3}$$

The error bound is approximately $\left|\epsilon_p\right|_{max} = \frac{M}{N}$ (4)

Thus, suppose that the maximum allowable instantaneous-period error is δ, then M must be restricted so as to lie in the range $0 \leq M \leq N\delta$. Figure 2 shows a plot of the two errors versus M/N, the scaling parameter.

The instantaneous-frequency varies between $\frac{1}{Q\tau}$ and $\frac{1}{(Q+1)\tau}$, whereas the desired frequency is $\frac{M}{N\tau}$, giving rise to errors of:

$$\epsilon_f(-) = \frac{N}{M(Q + 1)} - 1 \qquad \epsilon_f(+) = \frac{N}{MQ} - 1 \tag{5}$$

The error bounds are given by:

$$\left|\epsilon_f(-)\right|_{max} = \frac{1}{(1 + \frac{N}{M})} \qquad \left|\epsilon_f(+)\right|_{max} = \frac{1}{(\frac{N}{M} - 1)} \tag{6}$$

The nature of the errors are sketched in figure 3, while the error bounds are plotted in figure 4, allowing the usable scale range to be estimated for a given maximum error. Notice that for low scaling factors, the error bounds become M/N. The maximum instantaneous error in period or frequency can vary widely according to the precise scaling factor that is chosen, and positive and negative errors tend to vary reciprocally. Discrete points on the M/N scale where exact divisions are possible, are clearly shown in the graphs.

Although the average frequency is exact when measured over an interval of $N\tau$, practical applications do not usually allow such a fortunate choice of averaging period. Although the actual variation to be expected in an arbitrary assessment of frequency is difficult to determine, an upper bound can be placed on it. A scaled output frequency, generated after the style of the flow diagram in figure 1, has a cumulative error between the actual number of pulses in a given time and the desired number of pulses, of less than one. The maximum uncertainty in frequency measurement due to pulse irregularity, is then always less than 1/M. Notice that this is in addition to the normal digital measurement uncertainty of one pulse.[2]

2.2 Regular pulse period

An alternative strategy to that of figure 1, is to ignore remainders, thus rounding off the output periods to $Q\tau$ or $(Q + 1)\tau$, and giving a constant period. This type of scaling is sometimes to be preferred to that above, particularly in frequency synthesizer type applications, but there a loss of resolution is incurred.

The only output frequencies allowed are now of the form $1/Q\tau$, (Q integer), whereas the required frequency is $M/N\tau$. Thus the errors of figures 2, 3, 4 still apply, but the error is now in resolution or linearity rather than in instantaneous fluctuations. Provided that the scaling factor M/N is kept small, then the resolution errors are kept under control.

2.3 Use of a high frequency clock

When the input pulse-frequency is the only allowable time reference, the arrangement is only really suitable for small scaling factors. One way out of this impasse, is perhaps to make use of a reference clock of a much higher frequency than the input frequency, so as to provide a more finely divided scale.

Suppose that a reference clock is available, n times the input frequency, and that the output frequency is now expressed as a fraction of this higher frequency.

Thus: $$f_o = \frac{M}{Nn} \cdot f_c \qquad (f_c = n.f_i) \qquad (7)$$

This new situation is still describable by the previous calculations, except that only $\frac{1}{n}$ th of the clock frequency scale is used, so the worst errors are reduced by a factor of n.

The ideal situation considered in the paragraph above, may well in practical instrumentation, be compromised for convenience. For instance, Q periods of the input frequency may still be counted, but the remainder be interpreted by the new clock periods, as a multiple of τ/n.

Similar considerations apply when an arbitrary signal in pulse-frequency form is synchronised to a reference clock for further processing, variations occur in synchronised period which can be large unless the clock frequency is relatively high.

3. Some practical systems

3.1 Binary rate multiplier

The BRM has had a long and distinguished career,[3] is described in more detail for instance by Martin[1], and is now available in integrated-circuit form[4]. This device scales an input frequency by combining several of the pure division ratios. Output periods of the form $Q\tau$ and $(Q + 1)\tau$ are developed, but their spacing departs from that generated by the algorithm of figure 1, so that the error in an accumulated count is not limited to one pulse. Among several authors, Dunworth and Roche[5] give the maximum accumulated pulse error as:

$$\Delta = \pm \frac{1}{6}(n + 2.33) \qquad \text{(n is the number of BRM stages)} \qquad (8)$$

Consequently, the fluctuation in average frequency is greater for the BRM than for the ideal case considered above, although the fluctuations in instantaneous-frequency will be similar.

In all scalers of this kind, some reduction in frequency jitter can be achieved by use of a divider on the output pulse-frequency. Thus, a division ratio of P, results in a final output frequency $\frac{M}{N.P.\tau}$, but the fluctuation in period is reduced by P. This kind of trick can only be used where the initial frequency can be increased, for instance in a simple synthesizer.[6]

Using the BRM, a number of variations are possible, and Talbot and Senior[7] describe a simple system which scales a pulse train by a factor of 1/(1+a), ($a \leq 1$). An accumulated error of less than one pulse is claimed, but this is due to the small value of a which is used. It is characteristic of the BRM, that a small scaling fraction leads to improved performance.

3.2 Accumulator type scalers

Such a class of circuit seeks to implement with arithmetic logic the flow diagram of figure 1, or something similar. Butaev and Romashkan[8] describe such a device which is schematically illustrated in figure 5. An input code word of p bits is added into an n bit accumulator at every input pulse. An output pulse is produced when overflow occurs, and the accumulator is reset. The device thus produces pulses with period $(Q+1)\tau$, and incurs a linearity error of magnitude: $(1 + 2^{-n+p})^{-1}$. If the accumulator reset is suppressed, then the device operates with irregular pulse-periods, but a linear scale, as per section 2.1. As noted in section 2, the error may be made as small as desired in either case, by suitable restriction of the scale range. This circuit offers the ideal performance mentioned above, and is superior to the BRM in that the maximum error in accumulated pulse count is one pulse.

The Digital Differential Analyser (DDA) is also in this class of scaler.[1] Although using an accumulator operated by an input rate, in its conventional form there is no provision for setting in a digital code for accumulation, the register needs to be loaded serially when a new scaling fraction is required. The DDA element is available in integrated-circuit form, but is not very suitable for frequency scaling applications.

3.3 Frequency multiplication

The frequency scaling operations considered above, implicitly assume that the scaling factor is a fraction ($M<N$). Frequency multiplication is possible by using a frequency-tracking feedback loop,[1, 2] with a divider in the feedback path. A variable reference frequency f_r is established, which is divided by a ratio P before being balanced against the input frequency f_i by the feedback loop.

Thus: $$f_r = P.f_i \qquad (9)$$

In a feedback loop of this kind, the period irregularities become magnified by the frequency comparison and feedback, so the behaviour of the loop leads to further unknown factors. Techniques exist for the reduction of such irregularities, but these have been discussed elsewhere[1,3,9], although further work needs to be carried out in this area to determine the bounds of such fluctuations.

4. Conclusions

The theoretical behaviour of digital frequency scaling devices has been discussed, and it is shown that the maximum bound on instantaneous period and frequency errors is $\frac{M}{N}$, the scaling fraction. For low instantaneous errors, only the low part of the allowable scaling range can be used. The average-frequency fluctuations of a digitally scaled frequency are limited to 1/M. Frequency fluctuations can be eliminated, at the expense of scale linearity errors of M/N. These points have been illustrated by actual circuit devices, and the BRM has been shown to be non-optimum. Other scaling operations, such as frequency multiplication, squaring, etc., make use of feedback loops, which themselves require a separate treatment.

5. References

1. Martin, J.D., Signal Processing and Computation Using Pulse-Rate Techniques. The Radio and Electronic Engineer, 38, pp 329-344, December 1969.

2. Martin J.D., Digital Methods of Frequency Measurement: a comparison. The Radio and Electronic Engineer, 42, pp 285-294, June 1972.

3. Lundh, Y., Digital Techniques for Small Computations. J. Brit. Insts Radio Engrs., 9, pp 439-449, January 1959.

4. Texas Instruments Application Report B82. "SN 7497 Binary Rate Multiplier."

5. Dunworth A. and Roche, J.I., The Error Characteristics of the Binary Rate Multiplier. IEEE Trans. on Computers., pp 741-745, August 1969

6. Strauss, G.M., An AF Synthesiser. Electronic Design, 17, pp 62-65, August 16, 1971.

7. Talbot G.C.A. and Senior, R., Rescaling a Pulse Train. Electronic Engineering, pp 95-6, February 1968.

8. Butaev G.M. and Romashkan, V.S., Linear Code to Pulse-Frequency Converter. Measurement Techniques, 2, pp 194-196, February 1968.

9. Sumadi, Sosproprawiro. Computing Time Errors Introduced by Backlash Units. The Radio and Electronic Engineer, 42, pp 339-343 July 1972.

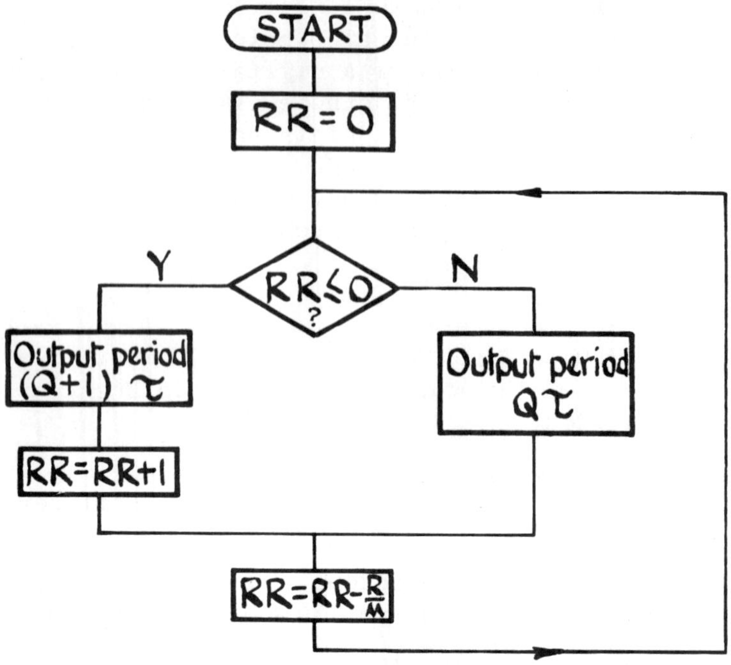

FIG.1. FREQUENCY SCALING ALGORITHM

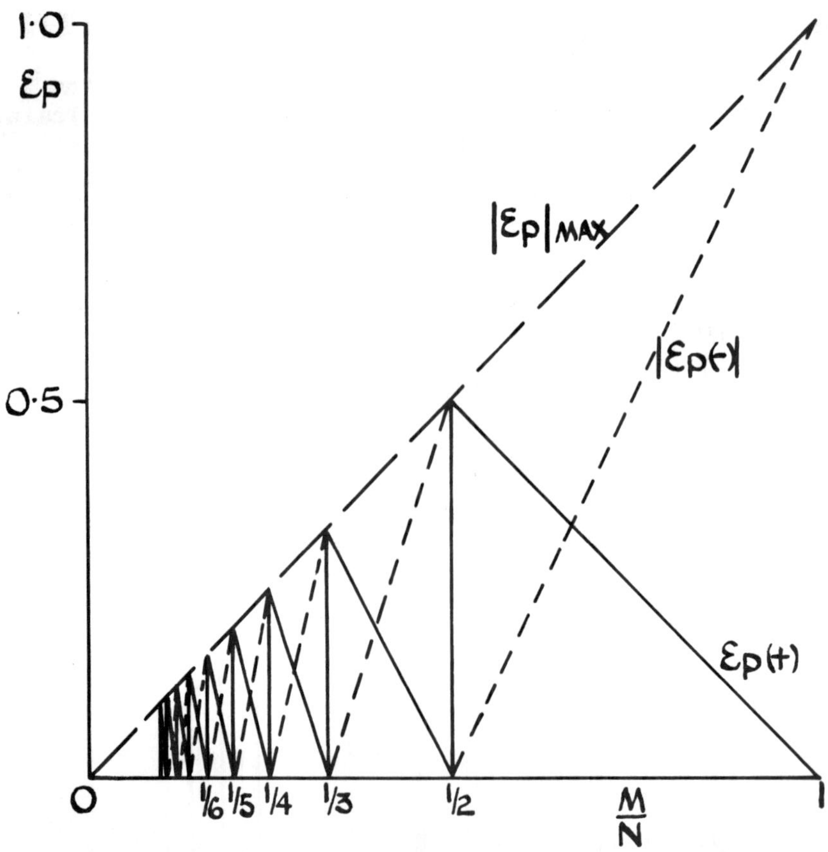

FIG.2. ERROR IN INSTANTANEOUS PERIOD

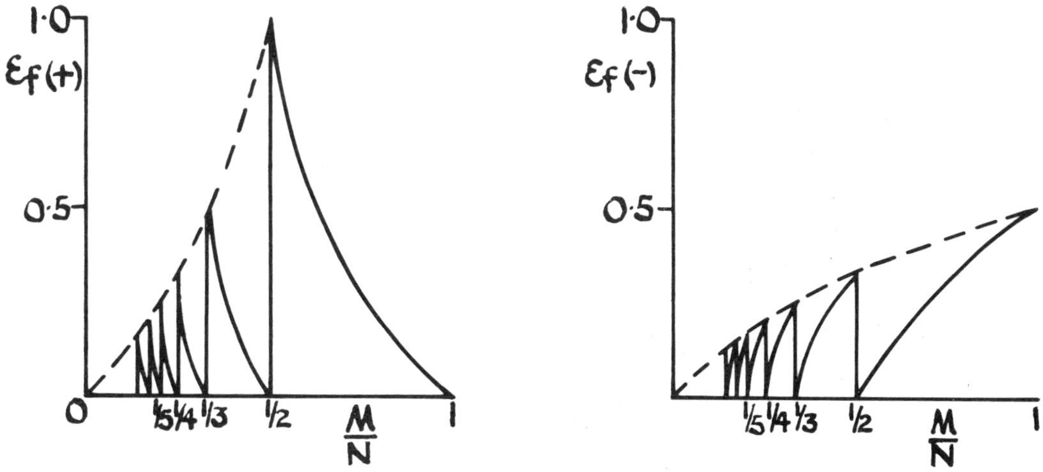

FIG. 3 ERROR IN INSTANTANEOUS FREQUENCY

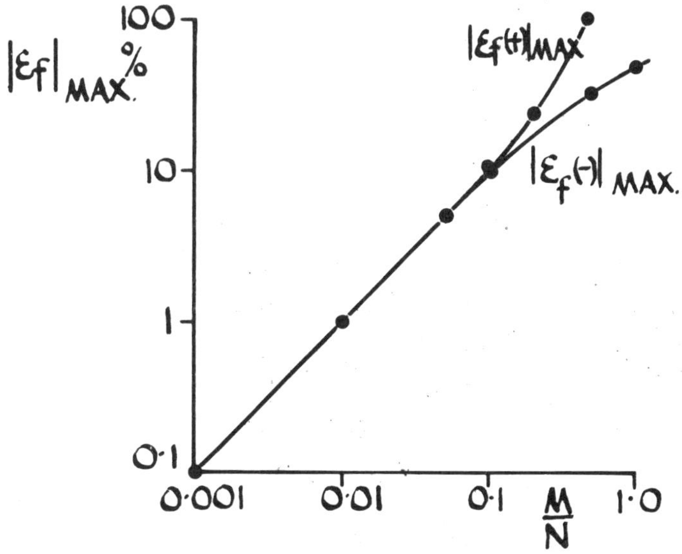

FIG. 4 ERROR BOUNDS FOR INSTANTANEOUS FREQUENCY

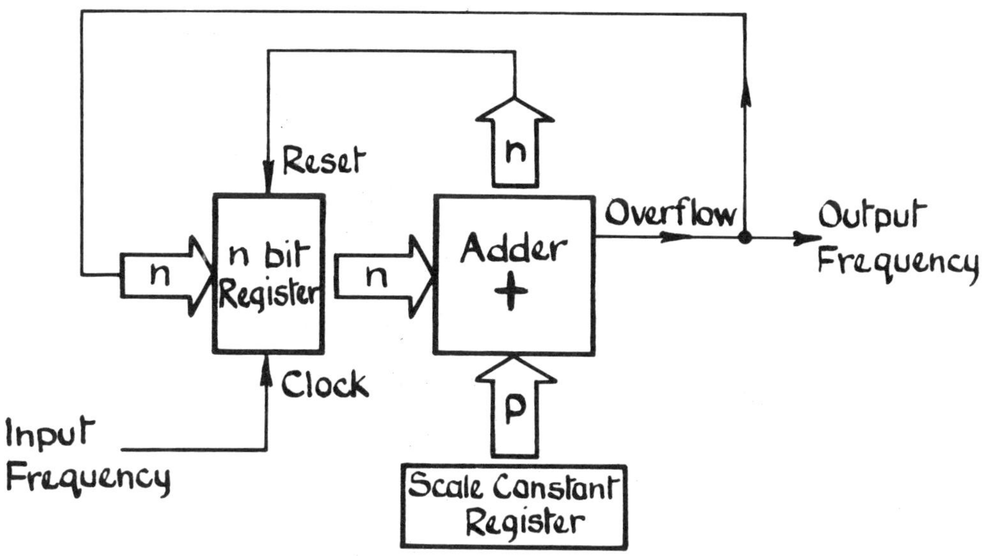

FIG. 5 ACCUMULATOR-TYPE SCALER

A PRECISION DIGITAL PHASE METER

R.N. Barnes and C.C.Williams

Introduction:

The digital phase meter described in this paper operates over a frequency range of 1.0 to 10 kHz with a resolution of 0.1° and up to 100 kHz with a resolution of 1.0°. The approach adopted overcomes many of the problems encountered with conventional phase meters such as uncertainty about accuracy over wide ranges of signal amplitude, the effects of distortion in the input signals, and the performance in the presence of noise. An additional serious problem also occurs with most electronic analogue phase meters when used at low frequencies due to the relatively long integration time (or time constant) required to achieve a steady reading. This problem also exists in those digital phase meters which are basically analogue in form, that is produce a d.c. voltage proportional to phase which is then displayed using a calibrated DVM.

Solution of the above mentioned problems has effectively been achieved by reducing the analogue type signal processing in the instrument to a minimum. In fact apart from two input comparators the entire instrument is digital in nature. Moreover careful selection of particular digital techniques has allowed relaxation of the specification of the input comparators so that the instrument is virtually free from drift. The techniques adopted allow substantially hands-off operation, all that is required for normal operation being to select the correct frequency decade, after which operation is automatic.

General Description

A simplified block diagram of the instrument is shown in Fig.1. (and relevant waveforms are shown in Fig.2.) The two input signals $\widetilde{A}$ (the reference) and $\widetilde{B}$ are fed through matched high pass filters, to isolate d.c. components in the inputs, to symmetrical limiting circuits. The resulting input signals then pass through high input impedance fast comparators yielding the waveforms A and B at their outputs. The edges of the A and B waveforms correspond to the zero crossover points of the input waveforms and ideally A and B will be square waves. In general however A and B will be rectangular due to drifts in comparator thresholds or to harmonic components in the input signals. The two signals A and B are logic signals and the EXCLUSIVE OR function $C = A \oplus B$ is formed as shown in Fig. 2(v). Now the average value of the C waveform is proportional to the phase shift and in an analogue phase meter the average would be derived by passing C through a low pass filter. At low frequencies the time constant of the filter would have to be large to obtain a steady voltage, consequently it would be better to derive the average value of C by integrating over a period of T_1 seconds where T_1 is an integral multiple of the periodic time of the

R.N. Barnes is with Thames Polytechnic.

C.C.Williams is with Feedback Instruments Limited.

input signals. In this case the phase is given by

$$\phi = \frac{1}{T_1} \int_0^{T_1} h\,dt = \frac{T_2}{T_1}.h \qquad (1)$$

To derive such a definite integral with a high degree of accuracy is difficult using analogue techniques, but a digital solution is possible. The waveform C is used to gate a high frequency clock, Fig. 2(vi) into a counter, thus if the total time for which the high frequency clock (at frequency F_c) is gated into the counter during one period T_1 of the reference signal is $2T_2$ (Fig.2(vii) then the number in the counter will be

$$X = 2\,T_2\,.F_c \qquad (2)$$

Clearly the count, X is proportional to the phase for a given signal frequency, but in order that the count should be proportional to phase at all frequencies, the result must be normalised by dividing by T_1 . The final indication will therefore be

$$= \frac{2\,T_2\,F_c}{T_1} \qquad (3)$$

McKinney[1] has described a digital phase meter where the normalisation is achieved by using a frequency ratio meter which basically measures the two times T_1 and T_2 and computes their ratio. However the approach adopted by the Authors here is to arrange that the clock frequency is always proportional to the reference signal frequency. That is

$$F_c = \frac{N}{T_1} \qquad (4)$$

The total count in a period T_1 vill be from (2) and (4),

$$X = \frac{2\,T_2\,N}{T_1} \qquad (5)$$

By choosing the multiplying factor N to be 1800 the 'count' will be

$$= \frac{3600\,T_2}{T_1} \qquad (6)$$

Consequently for a phase of 180.0° between A and B $T_2 = 1/2T_1$ and the display will be 1800. The resolution of the instrument is therefore 0.1°

A specially developed form of phase lock multiplier is used to generate the high frequency clock F_c and details are discussed in a later section.

The number displayed by the phase meter is the modulus of the phase shift in the range 0 to 180.0° and an auxiliary circuit is required to detect the sense of the phase. The B channel square wave may be inverted to allow for continuous measurement over the range -0.1° through 180° to $+0.1^{\circ}$, rather than the standard range of -179.9° through 0° to $+179.9^{\circ}$

The gating signal from the output of the EXCLUSIVE OR function consists of a sequence of pulses as shown in Fig.2. The selection circuit suppresses the inhibit gate preceding the counter until a "read" pulse is applied from the trigger circuit, causing the inhibit gate to open for the next complete cycle of A. Before the read pulse is applied the counter must of course be set to zero and all of these functions are perfomed by the trigger circuit. The phase measurement time thus equals one period of the input signal and this is particularly useful at low frequencies.

The correct functioningand hence accuracy of the instrument as a whole is dependent upon two basic processes, they are (i) the generation of the gate wavefom C shown in Fig. 2 and (ii) the proper generation of the high frequency clock. These two functions will now be discussed in detail.

Factors affecting the gate wavefom

The edges of the gate wavefom C correspond to the zero crossover points of the two input waves A and B and will be affected by (a) any offsets in the comparator thresholds (b) by the presence of hamonic components in either input signal and (c) any noise contamination of the input signals.

Threshold drift

The choice of an EXCLUSIVE OR gate for forming the C wavefom virtually eliminates the effects of threshold drift. The mechanism of the process is illustrated in Fig.3. The wavefom C_1 results when the B comparator threshold is at the zero level whilst C_2 results when the threshold has fallen to the level shown by the broken line on Fig. 3(ii). The total time for which the C_1 and C_2 wavefoms are at logic 1 are clearly equal showing that there will be no error due to such an offset. A limit to the correct ive action occurs when the offset voltage reaches a level given by $V_{os} = V_B \sin \emptyset$, where V_B is the peak amplitude of $\tilde{B}$ and $\emptyset$ is the actual phase between A and B. When the limit is reached the narrow pulse in the C_2 wavefom of Fig.3 vanishes; when the limit is exceeded the pulse reappears and constitutes an error. The rejection mechanism is also affected by hamonic components present in the input wavefom, but if the wavefom is affected in such a way that it remains symmetrical about its peak value (such as occurs, for example, when a sinusoid passes through a frequency independent transfer characteristic) then the corrective action still operates because the time advance in the positive edge of B_2 equals the delay in the negative edge.

The particular input arrangements adopted in the instrument afford an easy set-up procedure for the comparator thresholds. First the B input channel is grounded and a sinewave is applied to the A input. This causes the C wavefom to equal the A wavefom and the phase meter should consequently read 90.0°. The A comparator is therefore adjusted accordingly. Secondly the sine wave is fed to both inputs and the B channel comparator is adjusted to yield a phase reading of 0.0°.

Errors due to hamonic components

Nomally the phase of a distorted signal is considered to be the phase of the fundamental component, consequently the presence of hamonic components will in general affect the accuracy of phase meters dependent on the zero crossings of the input waveforms. McKinney[1] and Paull[2] outline some effects for certain types of "zero crossing" type phase meters, but a different approach is used here which leads to a full analysis of the effects of harmonics for this particular instrument.

The case is considered where one signal is pure and the other contains a harmonic component. The exact zero crossover instants may be found from the solutions of the following equations.

$$E_1 \operatorname{Sin} Wt + E_n \operatorname{Sin} n (Wt + \theta) = 0 \qquad (7)$$

where E_1 and E_n are the amplitudes of the fundamental and n_{th} harmonics, and θ is the relative phase of the harmonic to the fundamental. Relating the solutions of (7) to the effect on indicated phase error yields the following relations.

$$\phi_e = \frac{n(E_n)^2 \operatorname{Sin} 2n\theta}{2[E_1^2 - n^2 E_n^2 \operatorname{Cos}^2 n.\theta]} \qquad (8)$$

for n even and $E_n \ll E_1$ and ,

$$\phi_e = \frac{E_n \operatorname{Sin} n\theta}{E_1 - n E_n \operatorname{Cos} n\theta} \qquad (9)$$

for n odd and $E_n \ll E_1$

The worst case errors are therefore

$$\hat{\phi}_e = \frac{57.n E_n^2}{2 E_1^2} \qquad (10)$$

for n even and $2n\theta = \pi/2$

$$\hat{\phi}_e = \frac{57 E_n.}{E_1} \qquad (11)$$

for n odd and $n\theta = \pi/2$

The graphs plotted in Fig. 4 show the actual worst case errors (solid lines) together with the predicted errors (broken lines) from equations (10) and (11). The actual errors for low percentage distortions reflect the overall accuracy 0.2° of the instrument. Otherwise the agreement is good and creates confidence in the operation of the instrument.

It is not possible to predict in general terms errors due to 2 or more harmonics; each case must be treated individually. However if the total harmonic distortion on one channel is such as to cause the waveform to be symmetrical about its peak value there will generally be no error due to the rejection mechanism as described in the previous section.

Effects of noise

If one or both input signals are noisy, then the zero crossover points will be randomly affected, and random errors will occur in the phase readings. The effect becomes more noticeable as the zero crossover rate reduces, due either to a reduction of frequency or of input amplitude and in extreme circumstances noise can cause multiple transitions at the leading and trailing edges of wavefoms A and B in Fig.2. Whatever precautions may be taken to minimize the ingress of radiated noise or unwanted signals on power lines to the comparators, unavoidable circuit noise will eventually set a limit to the performance at low inputs and low frequencies or combinations of the two.

Under more normal conditions, where noise affects the crossovers only to the extent of producing random time variations in them, the variance of the phase readings (which will be dependent upon the signal to noise ratio) may be reduced by extracting the average of several readings. If M such readings are averaged, then the variance of the measured phase will be reduced by a factor $\sqrt{M}$. In the phase meter this is achieved by arranging that the selection circuit allows the inhibit gate to open for 10 complete cycles. At the same time a ÷ 10 stage is inserted between the inhibit gate and the counter, so that the count is the true average of 10 readings, and the variance will be reduced by 3.33 times. Phase measurement under these circumstances occupies a time equal to 10 periods of the input signal.

The phase lock loop

The phase lock loop must produce exactly 1800 clock pulses during a period of the input signal, and moreover the rate of these pulses must be uniform. A typical conventional phase lock loop is shown in Fig.5 and a comprehensive review of such loops has been given by Gardner.[3] Briefly, if the loop in Fig.5 is locked, i.e $F_o = F_a$ then the signal at the output of the Voltage controlled oscillator (VCO) will be at a frequency of 1800 F_a. Should the input frequency change then the d.c component in the phase sensitive detector (PSD) output will progressively change and will adjust the VCO to make F_o again equal F_a. However there is a problem in this application if a conventional phase lock loop is used, since the switching components present in the PSD output must be virtually removed by the filter otherwise the VCO output will be frequency modulated by residual ripple. The cut off frequency of the filter must therefore be well below the frequency of F_a and this in turn seriously restricts the dynamics of the locked loop. It means that the bandwidth of the closed loop phase control system will be very much less than the "carrier frequency" F_a, which can be as low as 1 Hz. It follows that such a loop will take a considerable time to follow a change in input frequency, and an even longer time to achieve lock in the first instance. Additionally the acquisition behaviour of conventional loops is such that they must be tuned generally to within an octave of the source frequency, and in this application it was felt to be desirable that lock acquisition should be rapid and automatic over the dynamic range of the VCO i.e. a frequency range of about 10:1. To meet these two requirements a new form of phase lock loop was developed, the essential features of which are shown in Fig. 6.

The PSD used is of a special form which acts as a phase detector when $F_a = F_o$ and as a quasi frequency discriminator when $F_a \neq F_o$. Consider the operation when $F_a \neq F_o$. Examination of the loop shows that it is effectively a Type 1 frequency control system and will therefore tend to approach zero frequency error. When this occurs the PSD acts as a phase detector, the loop is therefore a Type 1 phase control loop, and tends to a zero phase error condition (F_o in quadrature with F_a for a full wave PSD). The output of the PSD under these conditions will be a square wave at twice the frequency of F_a and consequently of F_o. The squarewave at the PSD output is removed by subtracting a signal at $2F_o$ derived from the 1800 divider chain. This form of loop reaches locked steady state conditions very rapidly as compared to conventional loops and has zero ripple at the VCO input. Thus the 1800 F_a frequency at the VCO output is at a perfectly uniform rate. These techniques ensure that no errors arise in the digital section of the instrument.

Conclusion

The design of a digital phase meter has been presented together with an analysis of some of the principal causes of errors. The main design philosophy was to produce an accurate easy to use instrument, and it is felt that the extensive use of digital techniques, carefully chosen with regard to the particular "A - D conversion" problems associated with phase measurement, has achieved the basic aim.

Acknowledgements

The authors wish to thank Mr.D.C.Carey for his contributions to the original conception of the instrument and Messrs. Feedback Instruments Ltd., who carried out the detailed development work leading to a production instrument.

References

1 J.E. McKinney. Digitised low frequency phasemeter assembled from logic modules. Journal of Research of the National Bureau of Standards Vol 71C No3 Jul - Sept 1967.

2 C.J.Paull. Phase measuring with increased accuracy Electronic Engineering July 1971.

3 F.M.Gardner. Phaselock Techniques. Wiley 1966.

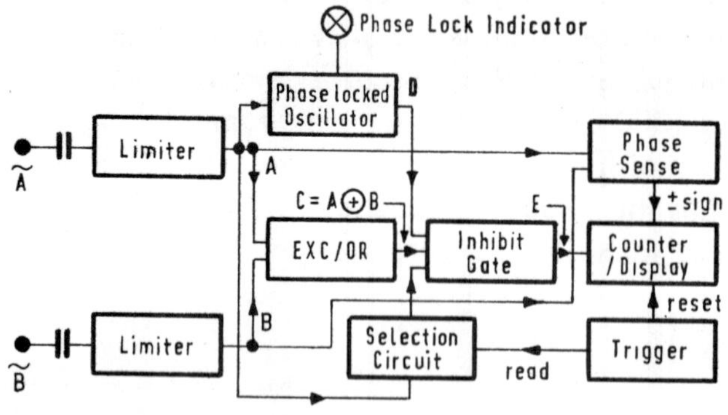

FIG. 1. BLOCK DIAGRAM

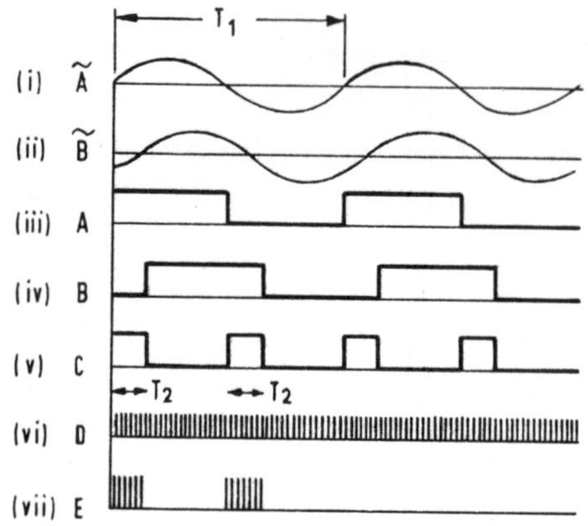

FIG. 2. WAVEFORMS

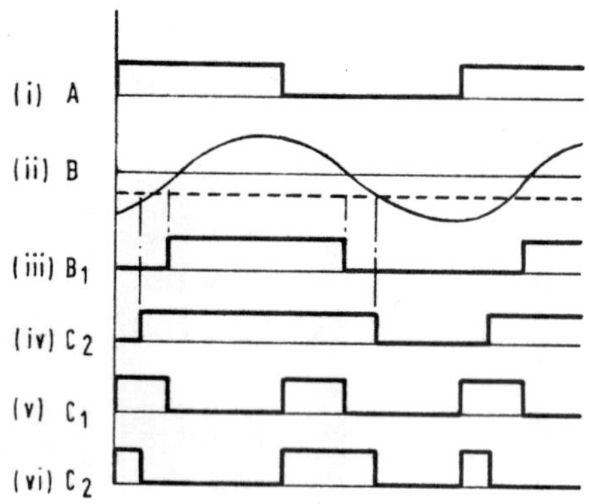

FIG. 3. EFFECTS OF COMPARATOR THRESHOLD DRIFT

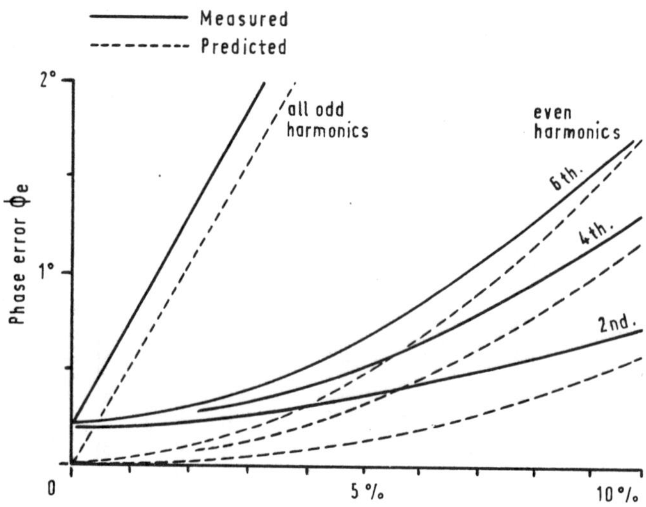

FIG. 4. WORST CASE ERRORS DUE TO HARMONIC COMPONENTS

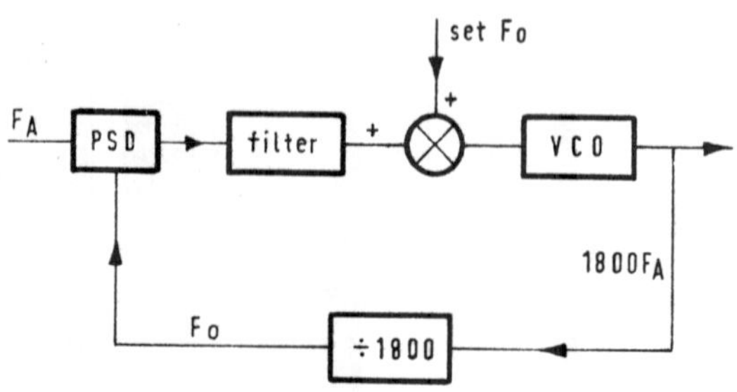

FIG. 5. CONVENTIONAL PHASE LOCK LOOP

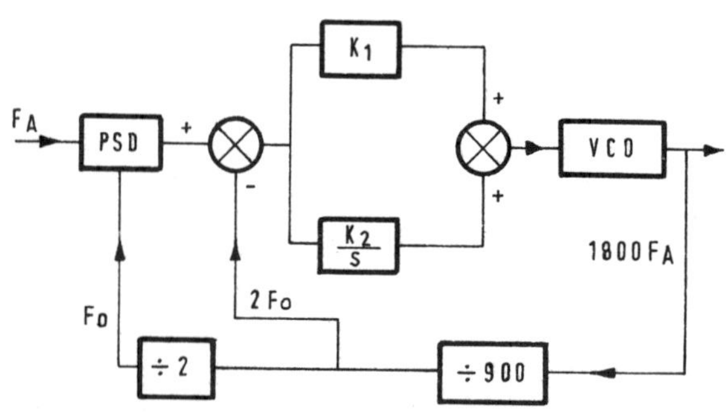

FIG. 6. MODIFIED PHASE LOCK LOOP

A NEW APPROACH TO DIGITAL METHODS OF CONVERTING MEASURED DATA TO ENGINEERING UNITS

R.J. Cooke

INTRODUCTION

For many years it has been generally accepted that a data logging system should be capable of providing an output not only of measured electrical quantities, but should have the facility of converting when necessary, to original engineering units. This requires an operation at some stage in the measurement cycle which compensates for transducer transfer characteristics and provides an output corresponding in value to the input stimulation in a desired unit of measurement. One method of achieving this is by analogue signal conditioning techniques but where voltage can be accurately measured and,a variety of sensors are likely to be employed, a convenient method is to digitally process each reading. Such an operation may involve a scaling factor with an offset but generally the transfer is non linear, e.g. thermocouples, and the approximating function more complex. The requirement is for essentially a look-up table and whereas this approach has its uses, the generally used technique has been to employ a series of tabulated points and to linearly interpolate between these.

If the general case of a continuous function which corresponds to the emf-to-stimulus relationship is taken for a transducer operating within a specified working range, then this may be expressed by means of a convergent power series of the form

$$y = a_0 + a_1 x + a_2 x^2 + a_3 x^3 + a_4 x^4 \qquad a_n x^n = \sum_{n-0}^{n=\infty} a_n x^n \qquad (1)$$

Where y is the stimulus, x is the output emf.

A truncated series will therefore describe a look-up table to a specified accuracy and the overall evaluation process is then one of approximation of a non-linear function by means of another. In order to limit the order of this equation it is useful to employ a different function for different sections of the output range and this is "idealised" by merely changing the values of the coefficients in an equation depending upon which section of the characteristic is being used. The remainder of this paper will be devoted to the derivation of these equations in conjunction with an economic hardware configuration for their implementation.

ALGORITHM APPROXIMATION

The nature of the problem of expressing the emf-to-engineering-units relationship for a transducer by means of a number of polynomials depends upon the transducer being used, the range over which measurements will be taken and the permitted deviation of the approximation from the tabulated values, Where there are definite discontinuities in the characteristic a separate low order polynomial will be required to convert each continuous range of emf's. The actual shape of the continuous curves is also important when attempting to fit with a low order polynomial as dominant high order coefficients will necessitate more interpolations and over smaller spreads of output voltage in order to achieve a desired fit.

In determining the number of zones required to describe a characteristic, such factors as the standardisation of the computing process for each zone and the total coefficient storage space are of vital importance; the latter the more so when the conversion is to be applied to a BCD data stream. It is therefore advantageous to approximate with only cubics, quartics or, in the simple case, straight lines and not to mix the different types of interpolating curves for one particular law. Indeed the zone lengths will most likely differ but the worst deviations from the curve in each zone should be about equal. In general the rms errors for each zone were equated when optimising a typical thermocouple curve.

The order of the polynomials used will increase as the degree of conformity is increased and as the zone lengths increase. For example consider a Chromel Alumel Thermocouple conforming to BS 1827 : 1952
The direct reading in °C must be within ± 1°C over the range -100 °C to 1350°C. A single cubic equation will suffice to convert emf's between 0mV and 25mV i.e. 0°C and 600°C but for the range -50°C to 600°C a fifth order equation will be required. Alternatively two cubics may be used. Cubic A from -100°C to 130°C and B from 130°C to 1350°C giving an overall fit of ± 0.9°C. The advantages of the latter approach is that a single computation of $a_0 + a_1 x + a_2 x^2 + a_3 x^3$ is simpler and quicker than its fifth or higher order counterpart. The additional comparison to determine which set of coefficients is required is a disadvantage for one law taken in isolation, but when a number of transducers are to be accommodated the added flexibility of the two zone method gives a good return on this increased complexity.

R.J. Cooke is with Solartron Schlumberger.

EXPRESSING A FUNCTION AS A SET OF POLYNOMIALS

The tabulated data for all the thermocouples to be used was assembled and each characteristic was explored over the required range using computer techniques to estimate the order of polynomial required to give a specific fit. In fact the tolerances on the approximation differed with each transducer but it was nevertheless found that two cubic equations adequately covered all the necessary ranges. At this stage it was also determined that the two zone method was the only way of achieving the specification if equations of 6th order and above were to be avoided. The study continued with a computerised optimisation of the parameters for each thermocouple law. Thus the two sets of four coefficients together with the point at which the curves crossed were adjusted until the overall rms error was minimal. Care was taken so that an even distribution of error occurred between the two zones by ensuring that each contained approximately the same number of reference points, taken from the published tables. Hence each law could be expressed as:

$$a_0 + a_1 x + a_2 x^2 + a_3 x^3 \qquad (2)$$

$$b_0 + b_1 x + b_2 x^2 + b_3 x^3 \qquad (3)$$

where (2) is the equation for $S \geqslant x$
(3) is the equation for $S \leqslant x$
the parameter S being the value of x for which

$$\sum_{n=0}^{n=3} a_n x^n = \sum_{n=0}^{n=3} b_n x^n \qquad (4)$$

From the above statement it follows that a preliminary comparison of x with S causes selection of coefficients a_n or b_n and thereafter the conversion is merely the evaluation of the corresponding cubic equation. In order to save computation time and to minimise the number of BCD working registers the equation (1) is best represented as

$$[\,(a_3 x + a_2)\,x + a_1\,]\,x + a_0 \qquad (5)$$

this being an add and multiply recursion.

AN ALGORITHM FOR MULTIPLICATION

Considering two registers containing A and B in BCD, then to form $A \div B$ we use register B as the master and operate on it until the contents are 1 and by applying the same operations to A the contents of this register become $A \div B$.

$$\text{Since} \quad B \times \phi(B) = 1$$

$$\frac{A}{B} \rightarrow \frac{A\,(\phi(B))}{B\,(\phi(B))} \rightarrow \frac{A/B}{1} \qquad (6)$$

In order to define ϕ it is necessary to consider B to lie within the range $0.1 \leqslant B < 1$. In practice this is easily achievable by applying an operator which ensures that the original value B is modified so as to conform with the above condition. Such an operator would, in the case of a BCD number, involve multiplication or division by powers of ten.

Once the position of the most significant digit is fixed, B is doubled repeatedly until $(2)^{n0}B$ just exceeds unity. This point is detected by testing the units decade for non-zero content. The register contents now undergo a subtraction of one tenth of B and this is repeated until the units decade becomes zero once again. Now the register contains $(2)^{n0}\frac{9}{10}^{n1}B$. By adding one hundredth of the register until unity is again reached the register contents become

$$(2)^{n0}\left(\frac{9}{10}\right)^{n1}\left(\frac{101}{100}\right)^{n2}B \;\simeq 1 \text{ here the error is less than } \frac{1}{100}\,B$$

Thus ϕB becomes

$$B\,(2)^{n0}\left(\frac{9}{10}\right)^{n1}\left(\frac{101}{100}\right)^{n2}\left(\frac{999}{1000}\right)^{n3}\left(\frac{10001}{10000}\right)^{n4} \ldots\ldots\ldots\ldots \simeq 1 \qquad (7)$$

When similar operations are applied to the contents of register A:

$$\frac{A}{B} = \frac{A(2)^{n0}\left(\frac{9}{10}\right)^{n1}\left(\frac{101}{100}\right)^{n2}\left(\frac{999}{1000}\right)^{n3}}{B(2)^{n0}\left(\frac{9}{10}\right)^{n1}\left(\frac{101}{100}\right)^{n2}\left(\frac{999}{1000}\right)^{n3}} \ldots\ldots \simeq \frac{A/B}{1} \qquad (8)$$

For multiplication two such sequences are required. The first to compute $\frac{1}{B}$ and the second to evaluate $A \div (\frac{1}{B}) = A \times B$.

In order to implement the algorithm simply, the second BCD register A is not in fact required as the control sequence may be applied successively to the same register. An ' auxiliary' register which stores four bit binary words is now used and the values n_0, n_1, n_2, n_3 which are the numbers of iterations per step are entered to it. The initial value B is loaded and by means of the above technique, is reduced to unity. The stored iterations are now used to process the register contents and to yield $\frac{1}{B}$ which subsequently undergoes a further series of operations to reduce its value to one. At this point, the stored data in the auxiliary register, when applied as indicated, will multiply the contents of the BCD register by the value B. If x is entered initially to B then X is generated such that $x = \frac{1}{X}$

$$\therefore a_3 x^3 + a_2 x^2 + a_1 x + a_0 \qquad (9)$$

is evaluated as

$$[\,(\frac{a_3}{X}) + a_2)\frac{1}{X} + a_1]\,\frac{1}{X} + a_0$$

$$\text{i.e. } \phi(\phi(\phi(a_3) + a_2) + a_1) + a_0 \qquad (10)$$

This is achieved by adding the coefficients $a_3\ a_2\ a_1\ a_0$ one at a time to the content of the register and applying the "operator" following each addition. A program to generate the function (9) above would be as follows:

Load x to the empty register	
Position and reduce to one.	This generates ϕ_1
Apply ϕ_1 to the contents of the register.	This generates $\frac{1}{x} = X$
Position and reduce to one.	Generates ϕ_2
Load a_3 to the empty register	
Apply ϕ_2	This generates $a_3 \div \frac{1}{X} = a_3\, x$
Load a_2 additively	$a_3\, x + a_2$
Apply ϕ_2	$(a_3\, x + a_2)\, x$
Load a_1 additively	$(a_3 x + a_2)\, x + a_1$
Apply ϕ_2	$(\ (a_3\, x + a_2)x + a_1)x$
Load a_0 additively	$(\ (a_3\, x + a_2)x + a_1)x + a_0$

Thus the transducer output x is initially processed to generate the control operator ϕ_2, which is used repeatedly to perform each multiplication by x. The arithmetic operations which lead to the generation of x and ϕ_2 take place in the BCD register and associated values of n_0, n_1, etc. are stored in the binary "auxiliary" register. This second register acts as a read only memory during subsequent stages when new data is entered from a coefficient store and is operated upon by the control operator. Finally when all the coefficients have been entered the BCD register contains the value of the desired polynomial. The entire program for the above operations may be stored in a ROM which, via a control logic block, conditions the registers and data ROM (coefficient store) and implements the operating sequence.

THE HARDWARE FOR A WORKING COMPUTER

The hardware required for implementing the aforementioned operations, may be conveniently divided into four main sections.

A) The computing register with its input and output data ports.

This BCD register, either Random Access Memory or Shift Register, accepts data from either an external source or the Read Only Memory which contains the coefficients for the polynomials and also the crossover point *S*. Further, data can be derived from the register itself via a selector which calls for the register contents multiplied by one, one tenth, one hundredth etc.

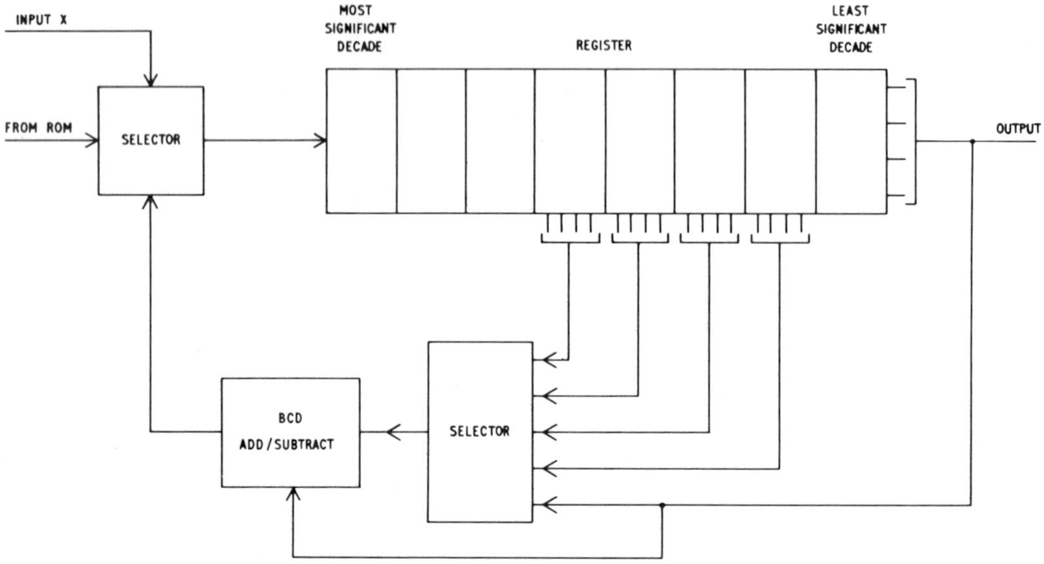

Fig. 1

B) The Auxiliary Register and Iteration Counter.

This is arranged in four bit words, each word being the number of iterations required to perform one stage of the control function. The contents of the register are loaded to the iteration counter when applying the control operator and the counter contents are stored in the register when generating the control operator.

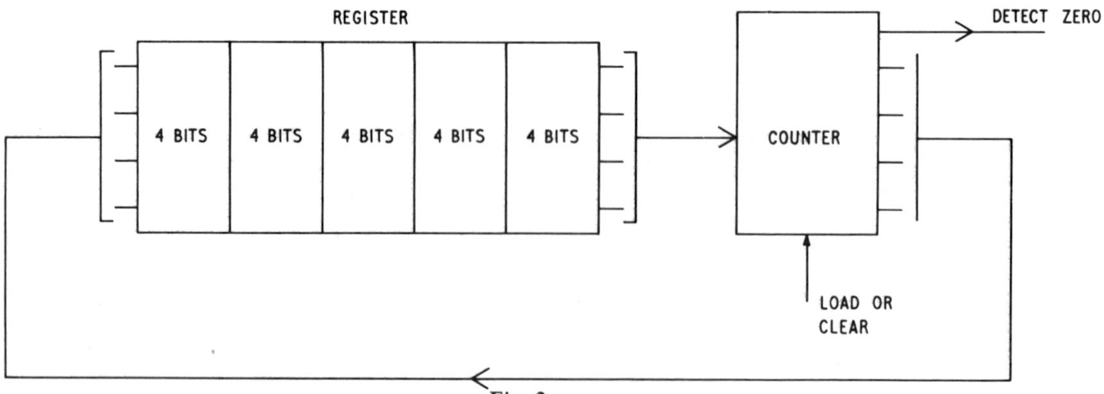

Fig. 2

C) The Data ROM

The data ROM contains the crossover point *S* and all the coefficients for the polynomials. The selector associated with the output is used to access the coefficients for either equation.

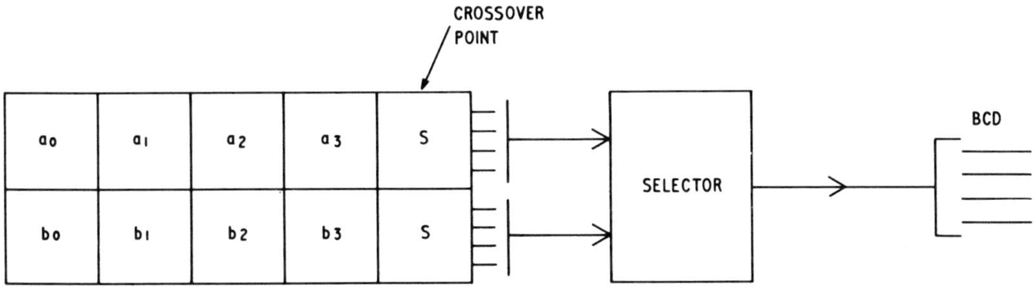

Fig. 3

D) The Control Logic

This is primarily a small ROM (256 bits) which is used to store the instruction codes of the overall programme; and the interpreting hardware, which drives the other components of the computer.

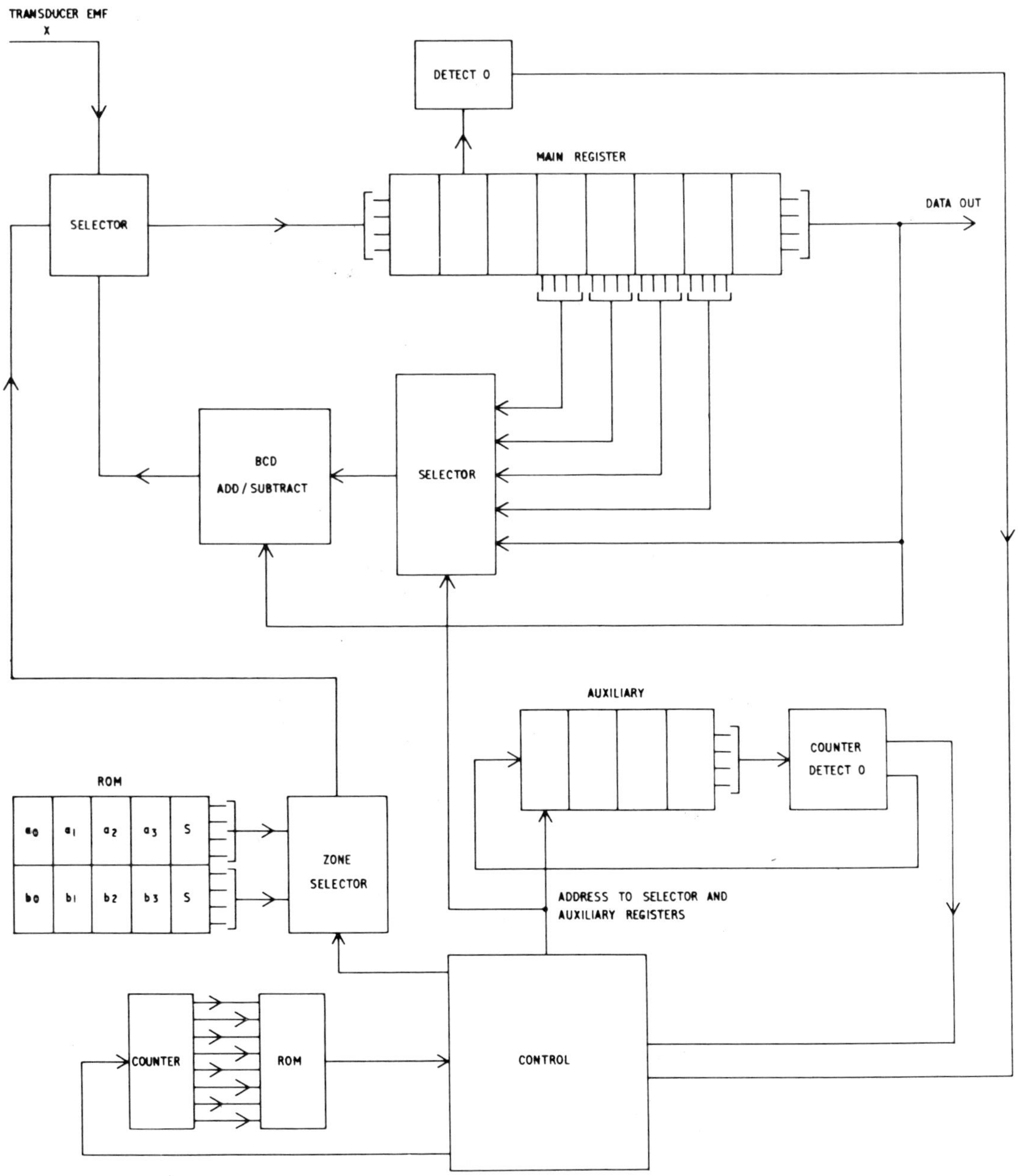

Fig. 4

EXAMPLES

1) Platinum - 13% - Rhodium v Platinum Thermocouple

The two cubic equations derived for the range 0°C to 1600°C to within ± 1°C of BS 1826 : 1952 using the IPTS–68 temperature scale yield the error curve illustrated in fig. 5.

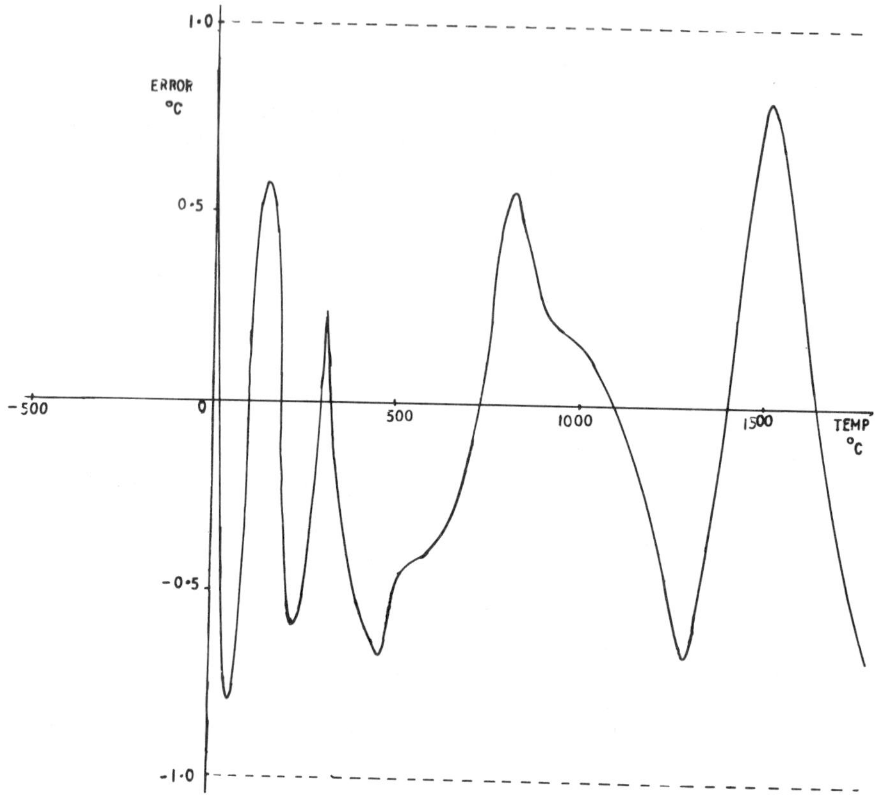

Fig. 5 Platinum - 13% - Rhodium v Platinum

Error vs temperature, for thermocouple linearisation

2) Platinum Resistance Thermometer

The two cubic equations which cover the range -200°C to 600°C to within .2°C (also giving -200°C to 900°C to within .3°C) of BS 1904:1964 (IPTS–68 temperature scale used) give the error curve of fig. 6. The output volts x are measured with a constant current passed through the resistance element.

The optimisation process may be varied to cause the error to be more or less confined to a particular range of temperatures and thus it is possible, as in the case of the Platinum Resistance Thermometer, to provide a closer approximation for a predetermined short range within the overall range covered by a particular lineariser law. The fit obtained using the polynomial technique compares favourably with a fifteen break point linear interpolation for all common thermocouple laws.

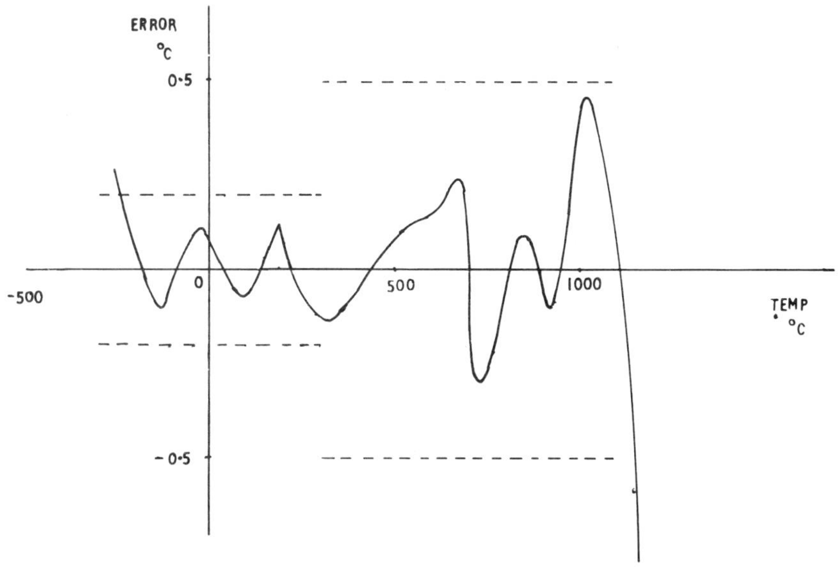

Fig. 6

Error vs Temperature for Platinum Resistance Thermometer Linearisation

ACKNOWLEDGEMENT

The author wishes to express his thanks to the directors of The Solartron Electronic Group for their permission to publish the information contained in this paper and for facilities made available for its production.

A DIGITAL SYSTEM FOR THE MEASUREMENT OF TRUE R.M.S. QUANTITIES

J. C. C. Nelson

Introduction

Although alternating electrical quantities are frequently referred to by their r.m.s. value, most measuring instruments respond to the mean (sometimes peak) value. Consequently the calibration of the instrument depends on the form factor (or crest factor in the case of a peak responding meter) of the input. The observed readings are, therefore, in error for input signals having a waveform different from that with which the instrument is calibrated (generally a sine wave). Instruments which indicate the r.m.s. of an alternating signal regardless of its waveform are of more general application and are known as 'true r.m.s. reading'. They are of use in measurements of noise and vibration and, in fact, in any system where non linear elements introduce harmonics.

This paper describes an instrument[1] designed to meet the following specification

i) the output should be a digital measure of the true r.m.s. of the input in a form suitable for data processing or direct display (i.e. as a digital voltmeter).

ii) the instrument should accept a high crest factor for a wide range of input signal levels.

iii) the response time should be as short as possible consistent with the need to determine an average value.

iv) if possible the need to compute square roots explicitly should be avoided.

v) a high accuracy should be maintained over the widest possible bandwidth.

Items ii) and iii) effectively rule out all forms of r.m.s. to d.c. conversion in conjunction with a conventional analogue to digital converter since a higher performance is possible if the r.m.s. measurement is an inherent part of the digitising process. Thermocouples, and related thermal devices[2], were rejected on the grounds of slow response, sensitivity to ambient temperature and vulnerability to overload.

Most commonly used analogue to digital converters may be broadly classified as integrating types or feedback types[3]. The latter make use of a digital to analogue converter, the output of which is compared with the input to be measured. A variety of strategies may be used to select the required digital number which brings the two analogue signals to equality. One of the simplest is to use a reversible counter with the direction of counting controlled by the sign of the difference between the analogue signals. The system to be described is based on a modification of this last method.

The author is with the Department of Electrical and Electronic Engineering, The University, Leeds LS2 9JT

Mode of Operation

A block diagram of the complete measuring system is given in the figure. The input signal, which will normally be an alternating voltage, feeds an operational amplifier, A, having manually switched input resistors which provide facilities for range selection and ensure that the signal to be handled is within the dynamic range of the remaining elements of the system. The range selection switch also determines the position of the decimal point in the display. The overall gain of the operational amplifier, for a given range, is accurately defined by a network of feedback resistors which can be switched in or out of circuit by means of logic signals.

The output of operational amplifier A feeds a device having a square-law transfer characteristic which in turn feeds a smoothing filter. The mean square signal generated by this filter is compared, by subtraction in a second operational amplifier, with a stable reference voltage. The sign of the difference, as detected by a comparator, controls the direction of operation of a reversible counter which in turn determines the feedback configuration of the controlled operational amplifier.

The counter operates until equality is achieved between the mean square signal and the reference voltage. In practice this form of 'bang-bang' control together with the lag inherent in the smoothing filter results in large limit cycle oscillations. This problem is successfully overcome by using a voltage controlled oscillator as a pulse source for the counter. The rate of counting is made proportional to the modulus of the error magnitude and the 'bang-bang' control becomes effectively proportional; as there is only one dominant lag the system is inherently stable.

Consider an input E to the operational amplifier A and assume that the counter is stationary at a value represented by a digital number D. The feedback conductance of amplifier A is arranged to be proportional to D and hence the amplifier output is proportional to E/D. The output of the square law device, therefore, is proportional to $(E/D)^2$. In the equilibrium condition, the value of D will be such that the output of the mean value filter is equal to the reference voltage R or

$$\overline{\left(\frac{E}{D}\right)^2} = R$$

where the bar indicates a mean value. Hence

$$D = \frac{1}{\sqrt{R}} \sqrt{\overline{E^2}} \;;$$

that is D is a digital measure of the r.m.s. value of E where the averaging period is determined by the nature of the mean value filter.

This mode of operation shows several useful features.

i) no explicit square root computation is required. The square law device, by virtue of its position in the loop, effectively performs this function.

ii) the location of the square law device within a feedback loop of this kind ensures a substantially constant signal level (varying only with crest factor). This enables efficient use to be made of the non-linear characteristic over a wide input dynamic range and is important if inputs having a high crest factor are to be acceptable.

iii) the time constant of the smoothing filter may be selected in relation to the lowest frequency present in the input. This means that r.m.s. measurements on high frequency waveforms are possible in only a few cycles of the input. This is not possible with thermal methods.

iv) although r.m.s. measurement is the primary aim, by-passing of the square law device results in a mean measuring instrument. Normally a device having a modulus characteristic would be required to provide mean of modulus as opposed to true mean.

v) the system possesses many features in common with feedback control systems and hence many of the techniques used to improve response time (e.g. error-derivative feedforward) may also be applicable here.

Circuit Realisation

The operating principle outlined in Section 2 is amenable to implementation using small and medium scale integrated circuits.

Operational amplifier A has parallel feedback resistors which are switched in or out of circuit by field effect transistors. Each digit of the reversible counter when at logic 1 is arranged to switch in a feedback resistor with conductance proportional to the weighting of the particular digit concerned. The feedback conductance G_f is therefore given by kD, where k is some constant, since parallel conductances add. If the input and feedback resistance of the amplifier are R_i and R_f respectively the closed loop gain is given by

$$\frac{R_f}{R_i} = \frac{1}{R_i G_f} = \frac{1}{R_i kD}$$

which is as required in the argument of Section 2.

The squaring operation is conveniently performed by means of an appropriately connected monolithic multiplier. It is important to realise that the absolute accuracy of the instrument is not simply related to the accuracy of the square-law characteristic. For an input waveform of any given type the instrument may be adjusted to give a correct r.m.s. reading regardless of the quality of the square law characteristic. A non-ideal square law will then introduce errors for other input waveforms which will normally be of a lower order than the absolute error of the squarer.

The filter is required to produce a continuous average of the squared input signal over some specified interval. To obtain an accurate r.m.s. value it is necessary to average either over an integral number of cycles or over a large number of cycles. The former would provide a faster response but unless a priori information of the input period is available, the latter must be used. It can be shown[4] that averaging over more than eight cycles of the input reduces the maximum error below 0.5%. For a fixed period of integration T a filter is required which has a rectangular impulse response of duration T and amplitude 1/T.

This filter is characterised by a transfer function

$$F(\omega) = e^{-\frac{1}{2}j\omega T} \frac{\sin \omega T/2}{\omega T/2}$$

and can be approximated but not precisely realised. In most practical applications it is not necessary for the period of integration to be defined with such precision and a simple lag type of filter is adequate.

Comparison of the filter output with the reference voltage is conveniently performed by a conventional operational amplifier, the output of which controls the direction of counting via an integrated comparator amplifier and also determines the operating frequency of the integrated controlled oscillator. It is important that the oscillator should maintain a low (ideally zero) output frequency for zero control input, but linearity of the control characteristic is of little importance.

The counting and display circuits are entirely conventional. For digital voltmeter applications binary coded decimal operation would normally be employed; for analogue-digital converter applications pure binary coding could be used with some simplification in the feedback network of operational amplifier A.

Conclusion

A system has been described which enables a digital measure of the r.m.s. (or mean) value of the input signal to be determined. An accuracy of ± 0.5% of full scale should be possible using normal techniques with the possibility of some improvement at greater expense. Bandwidth is limited primarily by the operational amplifier A; 100 kHz should be readily realisable with further extension using suitable circuits in conjunction with a wide-band amplifier.

Although the reversible counting technique is less sophisticated than, say, a successive approximation approach, response time is normally limited by the filter time constant required. This time constant can be changed to suit the input frequency permitting fast measurement of high frequency waveforms.

The availability of small and medium scale integrated circuits makes the systems easy to implement in a practical form.

References

1. British Patent No: 1,127,442.
2. Hermach, F.L.: Thermal Converters as A.C.-D.C. Transfer Standards for Current and Voltage Measurements at Audio Frequencies. Journal of Research, National Bureau of Standards, 48, p.121 (1952).
3. Schmid, H.: Electronic Analog/Digital Conversions, Van Nostrand Reinhold, 1970.
4. Nelson, J.C.C.: Ph.D. Thesis, University of Leeds, 1965.

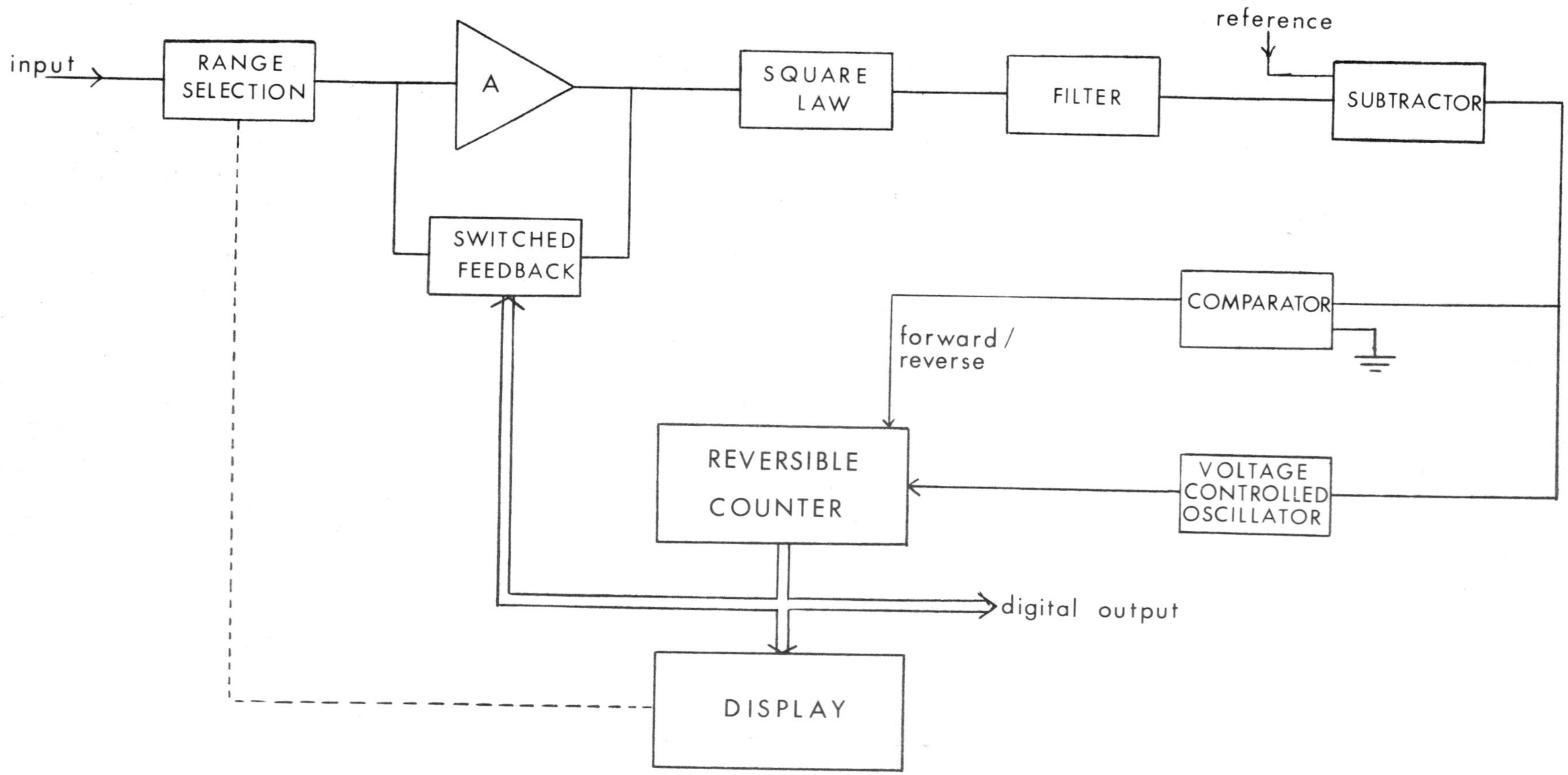

System Block Diagram

BINARY FLOATING POINT DIGITAL TO ANALOGUE CONVERSION

G.Hannington and D.G.Whitehead

Summary

This paper describes a binary floating point digital to analogue converter suitable for a graphics display system. Normalised floating point numbers are used and are converted in two simultaneous operations taking the mantissa and exponent separately. The exponent bits control the multiplication factor of an analogue multiplier. This multiplier obtains its input from a standard D/A R-2R ladder network controlled by the mantissa value. The system is capable of operating within a number range of 10^5 and has a settling time within 0.5 microseconds.

Introduction

A graphic display system, the centre of which is a specialised digital computer which operates as a complex function generator, is currently being developed at Hull. Floating point number representation is necessary due to the large number range involved and inherent scaling problems within the machine. In fact, in order to handle large number ranges, some form of floating point representation is used in most digital systems. A perennial problem encountered in the display of digital information on a C.R.T. is the difficulty of converting the information into an analogue form. The problem is exacerbated if the digital system operates in a floating point notation and for display purposes data is usually converted into analogue form thus:

$$\text{Floating Point} \xrightarrow[\text{software}]{\text{hardware}} \text{Integer} \xrightarrow{\text{D/A converter}} \text{Analogue} \longrightarrow \text{Display}$$

Conversion of floating point to integer representation by software routine involves the shifting of the mantissa left or right a number of places specified by the exponent. The hardware method mechanises the process and it can be significantly faster but must still take an appreciable time to perform (repeated shifting and comparison operations). Owing to the wide number range implicit in floating point representation, at this stage of the transformation a large capacity integer register is required, which in turn demands a long and therefore expensive integer D/A converter. The floating point to analogue converter to be described in this paper allows the conversion to be carried out speedily in a single operation without an intervening integer stage.

Any output system must inevitably make restrictions upon data to be displayed. In the case of a graphics display some important restrictions are the resolution of the screen and the deflection system. In fact some form of scaling of the data will always be necessary. An advantage of operating in floating point is that if, using the smallest exponent, the mantissa has sufficient resolution for the entire screen then scaling can be conveniently carried out by straightforward adjustment of the exponent.

Department of Electronic Engineering,
University of Hull.

The Converter

Table 1 shows the standard form of biased normalised floating point representation used.

Mantissa	Exponent	Decimal equivalent
0.11111111111	1111	approx. $+1 \times 10^{4\cdot5}$
0.10000000000	0000	$+\frac{1}{2} \times 10^{0}$
0.00000000000	0000	zero
1.11111111111	0000	approx. zero
1.01111111111	0000	approx. $-\frac{1}{2} \times 10^{0}$
1.00000000000	1111	$-1 \times 10^{4\cdot5}$

For positive numbers, $N = \left(2^{-1} + \sum_{i=2}^{m-1} B_i . 2^{-i}\right) \times 2^{-\sum_{j=0}^{e-1} B_j . 2^{+j}}$

where m = no. mantissa bits
e = no. exponent bits
and B_i, B_j have the value 0 or 1

Table 1.

Ones complement method is used in the mantissa to represent negative numbers. The total number range is approximately 10^5. Note that because the exponent is considered to be negative, all numbers are fractional.

Basically the scheme consists of a 12-bit D/A converter using a R-2R ladder network controlled by the mantissa, followed by a multiplier section whose multiplication factor is determined by the exponent.

1. The Exponent Multiplier

The simplest way of introducing the exponent is to use a series of switched feedback amplifiers, one for each exponent bit such that the overall gain is 2^n, where n is the value of the exponent (see figure 1).

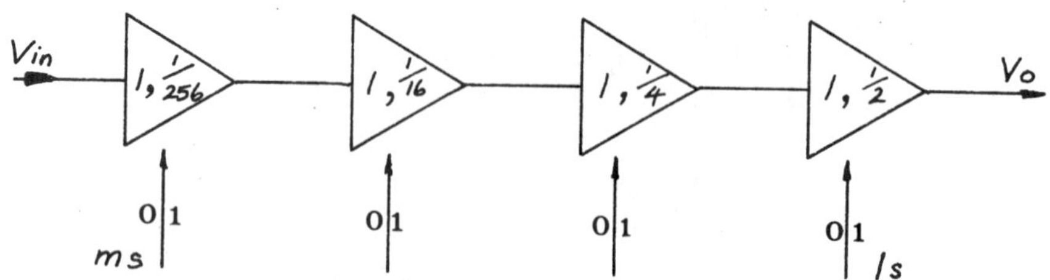

$$V_o = V_{in} . \left(2^{-\sum_{j=0}^{e-1} B_j . 2^{+j}}\right)$$

Figure 1

This arrangement can be reduced to two stages, each handling 2 bits. If it is arranged that the most significant and least significant bits are handled by the same amplifier, then the ranges of component values required are minimal. As mentioned earlier a biased exponent is used, this bias is of half the range of the exponent and obviates the necessity of a code conversion when dealing with negative exponents. The bias may be removed by a fixed gain amplifier or suitable gain adjustment in the display console. It is of practical benefit to define the exponent as negative as this allows the voltage swing on the mantissa to be large, giving a good signal to noise ratio at the exponent amplifier input over the entire number range.

2. The Mantissa Converter

As shown in Table 1, a 12 bit mantissa is used, and as is usual practice the most significant bit is taken as the sign bit. Due to the normalisation process, the next most significant bit is always different from the most significant bit, except in the special case of zero. A standard D/A converter network can be used here. In our case an R-2R ladder switched by TTL logic provided adequate resolution. Negative numbers can be converted directly to their analogue form in this system if they are represented by their complement. Two ways of achieving this are shown in Figure 2.

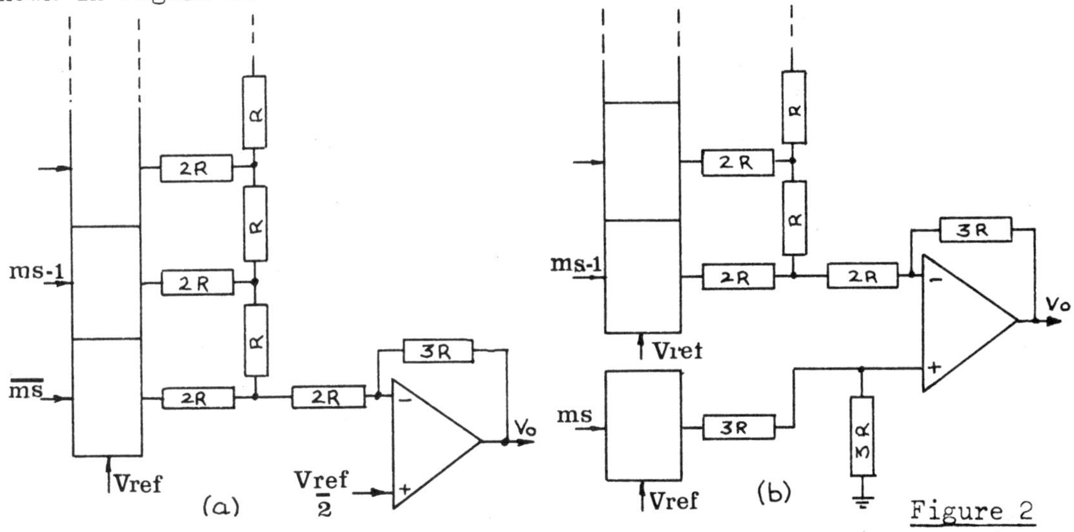

Figure 2

In both cases the output is bipolar. Note that the system is correct for truly represented negative numbers, that is twos complement. Ones complement representation can produce an error equal to the least significant bit. The circuit in Figure 2b, though requiring a ladder shorter by one bit than that in Figure 2a, suffers from circuit limitations inherent in the operational amplifier; due to its lower slew rate, changes on the non-inverting input considerably increase the amplifier settling time. This is compounded by the fact that this input also has the largest step.

The Complete Circuit

The full circuit is shown in Figure 3. The exponent amplifier gains are controlled by transistors T1 to T4. These switch appropriate gain controlling resistors in the feedback loops. It is important that the 'on' resistance of the FETs is sufficiently low so as not to affect the operation of the circuit; alternatively, this resistance may be taken into account at the design stage. An overall tolerance of 2% of f.s.d. was adequate for our display system. This demands that each multiplier amplifier must be accurate to within 0.67% (assuming the same accuracy for the mantissa D/A converter). This can be achieved quite easily. A sample-hold circuit is included for the suppression of switching transients and overshoot. The inclusion of compensation networks on the amplifiers made it possible to keep the transients to within 0.5 microseconds. As an alternative to the sample-hold circuit, Z-modulation of the display tube could be used.

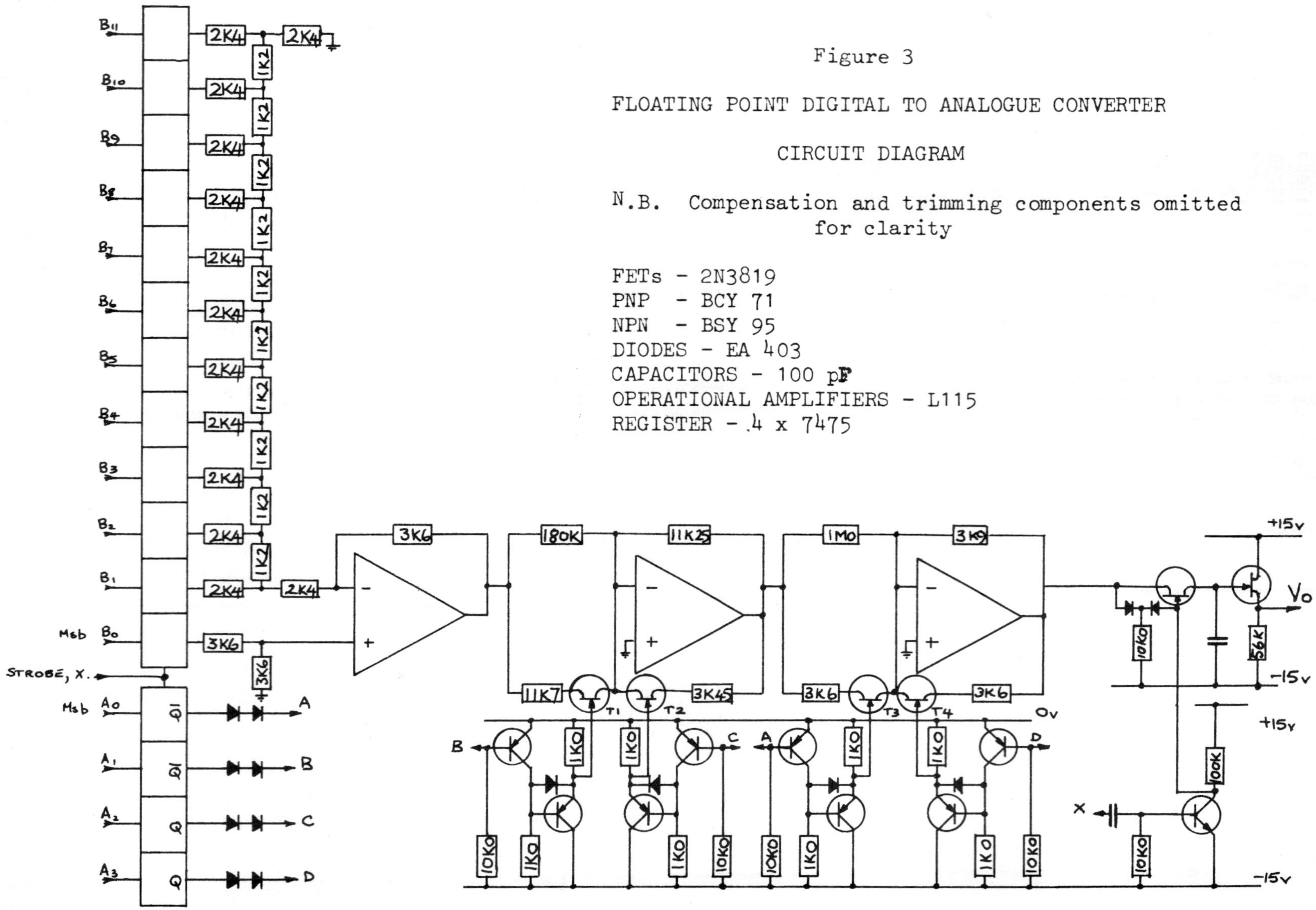

Figure 3

FLOATING POINT DIGITAL TO ANALOGUE CONVERTER

CIRCUIT DIAGRAM

N.B. Compensation and trimming components omitted for clarity

FETs - 2N3819
PNP - BCY 71
NPN - BSY 95
DIODES - EA 403
CAPACITORS - 100 pF
OPERATIONAL AMPLIFIERS - L115
REGISTER - 4 x 7475

The operation of the system is shown in figures 4 to 6. The top trace of figure 4 shows the output from the exponent section of the converter when switched between two gain states (2^0 and 2^{-1}) by the least significant bit. The input to this section is the square wave shown on the middle trace. Switching transients can be seen on the edges of the output steps. Figure 5 shows the separate operation of the mantissa and exponent sections when controlled by an increasing floating point count. The upper trace is of the exponent section with the mantissa input replaced by a fixed voltage level. Since the input is a floating point count which increases logarithmically with time, the output from this circuit when mantissa and exponent are connected together must be a linearly increasing voltage. Figure 6 shows such a response.

Conclusion

The system described has been built, tested, and in operation for several months. As we require only two of these circuits discrete component construction was employed and resistors trimmed to the required values. Since the gains of the exponent amplifiers are controlled by resistor ratios, it would obviously be possible and advantageous to use precision monolithic techniques here.

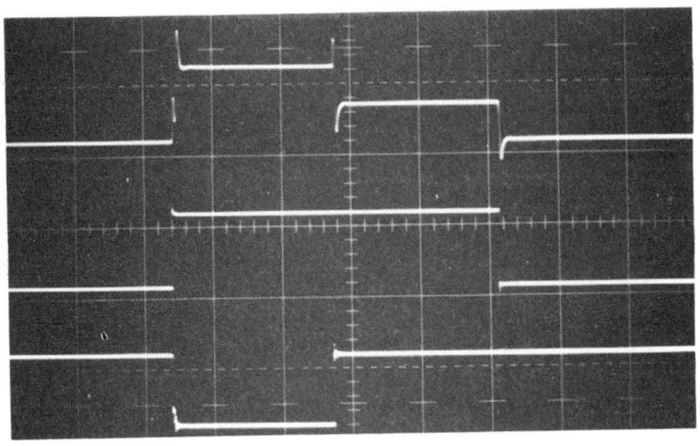

output

input

control (-2)

scale: 1v/div.
5μSec./div.

Figure 4

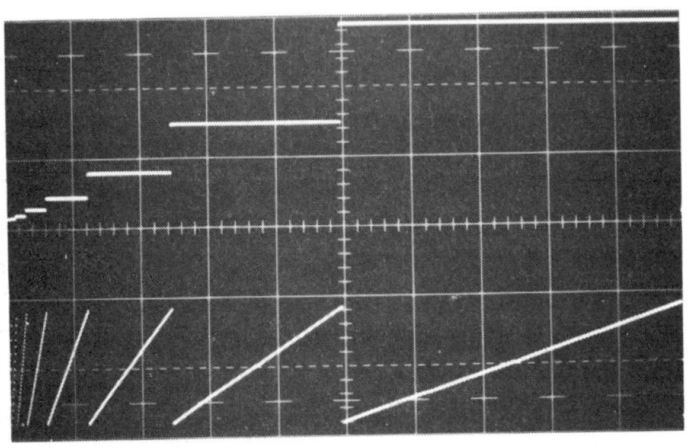

exponent

mantissa

scale: 1v/div.
2mSec./div.

Figure 5

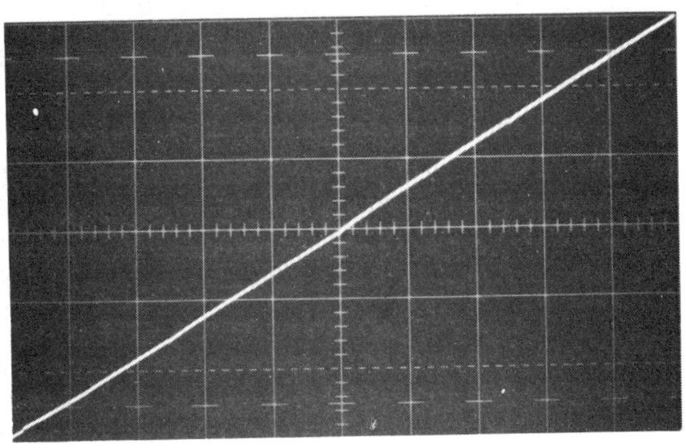

output ramp

scale: 0.5v/div.
2mSec./div.

Figure 6

AN ELECTROMAGNETIC DEVICE FOR TWO-DIMENSIONAL POSITION ENCODER

S. Inokuchi, K. Tokura and Y. Sakurai

Introduction

Recent works on two-dimensional position encoder for graphic data input systems of digital computer have resulted in the development of ingeneous devices such as the light pen, the RAND Tablet(1), the Sonic pen(2), the Hologram Tablet(3), and the Sylvania Tablet(4). These devices allow a user to perform a two-dimensional information input function, which cannot be conveniently handled by a keyboard. The graphic input systems are coming into increasing use for Computer Assisted Instruction, Computer Aided Design and the transmission of two-dimensional information through the telephone line. These devices, which are named by data tablet or digitizer, should be simple and easy to operate and to be inexpensive.

Electromagnetic induction is also applied to converting graphical data into a machine-readable signal(5,6). Electromagnetic device has the advantages as follows. The low impedance of the sensing device decreases the influences of noise and the accuracy is not affected by the variation of atmosphere such as temperature change. The devices described in the references (5) and (6), however, do not seem to be satisfactory, because the former requires a large number of sense amplifiers and the latter is attended with the writing pen including a magnetic head which must be held in the correct orientation.

The authors have developed an induction-type graphic data tablet(7). It consists of an electromagnetic stylus which generates alternating magnetic flux and sensing coils on the input tablet. This device requires a plurality of sense amplifiers, whose number is the same as that of division bits along X and Y axes.

This paper describes a new type of two-dimensional position encoder based on electromagnetic coupling of closed-loop exciting coils encoding the detecting plane with a sense coil wound upon a magnetic needle. As the plane and the stylus are used as the signal generator and the sensor, respectively, the number of sense amplifiers is only one and the greater part of electronics can be shared with a plurality of planes. The new device satisfies the following conditions: (1) simple and easy to operate, (2) high speed, (3) high signal-to-noise ratio and (4) inexpensive.

Principle

Figure 1 shows a closed-loop exciting coil on a detecting plane and a stylus including a sense coil wound upon a magnetic needle. The exciting coil divides the plane into the inner regions and the outer ones. If a rectangular current pulse is applied to the exciting coil, the electromotive force is induced in the sense coil of the stylus. The polarity of the e.m.f. of the stylus located in the outer region is opposite from that in the inner region, because the direction of the magnetic field inside the loop is different from that outside the loop. This makes it possible

S. Inokuchi, K. Tokura and Y. Sakurai are with Faculty of Engineering Science, Osaka University, Japan.

to know whether the stylus is positioned inside the exciting coil or not. By expanding this principle, it is possible to digitize the plane with a plurality of exciting coils fed sequentially with current pulses, and thus to encode the stylus position.

Input Plane

Figure 2 illustrates two-orthogonal sets of three-bit exciting coils arranged in X and Y axis directions on the detecting plane. Each coordinate is divided into eight regions 1 through 8 by three exciting coils. By considering the inner region of each coil as "1" and the outer region as "0", the plane is encoded into Gray code as shown in Table 1. The number of exciting coils required to resolve eight regions is only three. Accordingly, 2^n regions can be resolved by n exciting coils.

The Gray code was selected so that only one bit could change between adjacent regions, giving a complete and unambiguous determination of the stylus position. Furthermore, the number of exciting coils which divide adjacent regions is always only one. Thus, a plurality of exciting coils in each direction can be formed into one layer on the effective area of the detecting plane, whereby the fabrication of the plane is facilitated.

Circuit Configuration

The function of this device is illustrated by the timing diagram during a major cycle as shown in Fig. 3. Figure 4 is the block diagram of this system, where the heavy lines indicate the flow paths of the position information. The clock pulse generator furnishes a sequence of 18 pulses to the distributor, which applies rectangular current pulses p_1 through p_8 and p_{10} through p_{17} to the exciting coils in order of x_1, x_2, ..., x_8 and y_1, y_2, ..., y_8, and generates control pulses p_9 and p_{18}. The position information is sensed by the stylus coil (A) and applied to the polarity detector. The polarity detector outputs a "1" logic pulse when the stylus is located inside the corresponding exciting coil (B). Its output is serial Gray-coded signal, which can be converted to serial binary code by a J-K flip-flop (C). The position information, after being converted to parallel binary code by a shift register, is transfered to X and Y buffer storages for the output by control pulses.

Polarity Detection

The exciting coils and the sense coil are inductances and there exist mutual inductances between every two of them. So the current may be oscillatory when one of these coils is driven by a voltage pulse. In such a case it is impossible to detect the polarity of the e.m.f. by only detecting whether the positive pulse is induced or not during the driving pulse. In this device, the polarity of the e.m.f. is detected by means of the circuit shown in Fig. 5(a). Figure 5(b) shows the waveforms in its response. Comparator I gives a "1" level while the input signal is higher than a certain positive threshold level indicated by $+v_t$. Comparator II gives a "1" level while the input is lower than a negative threshold $-v_t$. This circuit can determine whether the positive leading end procedes or not during the driving pulse. Figure 6 is oscilloscope patterns of the signals in the polarity detecting circuit.

Prototype Device

In the prototype device, the tablet size is 256×256 mm^2 and the resolution is 1 line/mm. As the pulse sequence has 100 μsec in major cycle and the position information is detected once during one major cycle, the data

rate is 10,000 positions/sec. The exciting coils are driven by rectangularcurrent pulses of 0.2 A.
The magnetic sense stylus includes a straight slender magnetic needle (0.15 mm in diameter and 5 mm in length) made of permaloy wire. It has a sense winding of 100 turns at the tip. Permaloy was selected so that the induced e.m.f. could be large because of its high permeability.
Figure 7 shows the prototype and an example of monitored pattern.

Discussion

The plane size and the resolution depend upon the induced e.m.f.. The e.m.f. is proportional to the time-derivative of the magnetic field at the position of the sense coil. The magnetic field intensity changes extensively with the location on the detecting plane. Assuming now that the center of the sense coil is at the point P in Fig. 8(a), where h is the height above the exciting coil plane and d is the horizontal component of the distance from the coil wire. By the Biot-Savart law, the contribution of current element $I \cdot dl$ to the normal magnetic field component at P is:

$$dH = \frac{1}{4\pi} \cdot \frac{I \cdot dl \cdot \sin\theta \cdot \cos\phi}{l^2+h^2+d^2},$$

where $\sin\theta = \frac{\sqrt{h^2+d^2}}{\sqrt{l^2+h^2+d^2}}$ and $\cos\phi = \frac{d}{\sqrt{h^2+d^2}}$.

When integrated, this leads to:

$$H = \frac{I \cdot d}{4\pi(h^2+d^2)}\left[\frac{l}{\sqrt{l^2+h^2+d^2}}\right] \quad l\text{:entire length of exciting coil.}$$

Figure 8(b) shows the magnetic field at the center of one of teeth-like loops of the exciting coil near the center of the detecting plane. The center of the loop was selected because the field intensity becomes minimum at the center of the loop for wide loop coil. The field intensity decreases abruptly when the width W becomes smaller than the height h. As it is found from the figure that it is possible to resolve the loop coil whose width is one half of the height, the loop coil of 0.5 mm in width could be resolved in the case of h = 1 mm.
The plane size is twice the most significant coil width and the minimum width to be resolved is one half of the width of the least significant coil, as easily assented from Table 1. Thus, the plane size 1,024mm x 1,024 mm and the resolution 0.25 mm would be obtained. The resolution can be improved by using an alternative redundant unit distance code as shown in Table 2. It is found from the table that pitch of division is one fourth of the width of the least significant coil.

The data rate can be increased by raising the frequency of exciting current pulse. For this purpose the time constant L/R in the driving circuit must be decreased by increasing R, so that the transient response may die out within the duration of the exciting pulse. This problem would be solved through the utilization of a constant current pulse source. But from the view point of applications, the present rate is already sufficient.

Conclusion

The utilization of inductive coupling is useful in this application for several reasons: First, the low impedance of sensing device increases signal-to-noise ratio. Second, the accuracy is not affected by the variation of atmosphere such as temperature change. Third, an illustrated sheet can be placed over the plane to encode the illustration by manual tracing. As the detection of the induced e.m.f. is inherently considered as digital device, this system should further increase the effective reliability.

References

(1) M.R. Davis and T.O. Ellis, "The RAND Tablet: A man-machine graphical communication devices," Proc. 1964 FJCC, pp.325-331.
(2) A.E. Brenner and P. de Bruyne, "A Sonic Pen: A digital stylus system," IEEE Trans. Comput., vol.C-19, Jun. 1970, pp546-548.
(3) M. Sakaguchi and N. Nishida, "The Hologram Tablet- A new input device," Proc. 1970 FJCC, pp.653-658.
(4) J.F. Teixeira and R.P. Sallen, "The Sylvania Data Tablet: A new approach to graphic data input," Proc. 1968 SJCC, pp.315-321.
(5) U.S. Patent No. 3,598,903.
(6) M.H. Lewin, "A magnetic device for computer graphic input," Proc. 1965 FJCC, pp.831-838.
(7) S. Inokuchi, K. Tokura and Y. Sakurai, "A New Magnetic Device for Graphic Input," IECE Trans. vol.56-D, No.2, 1973.

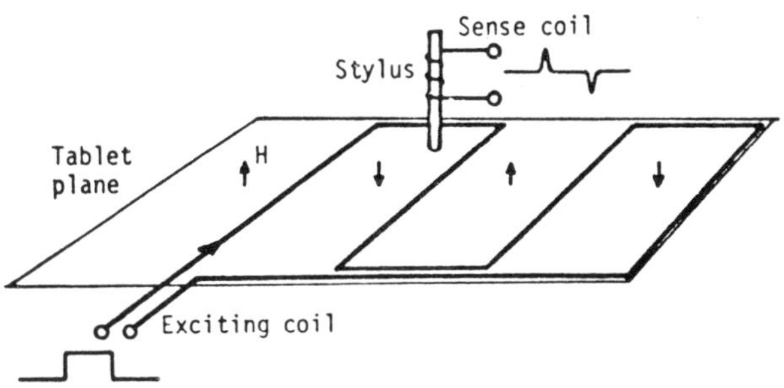

Fig. 1. Closed-loop exciting coil on tablet plane and sense stylus.

Region	X_1	X_2	X_3
①	0	0	0
②	0	0	1
③	0	1	1
④	0	1	0
⑤	1	1	0
⑥	1	1	1
⑦	1	0	1
⑧	1	0	0

Table 1. Gray code table

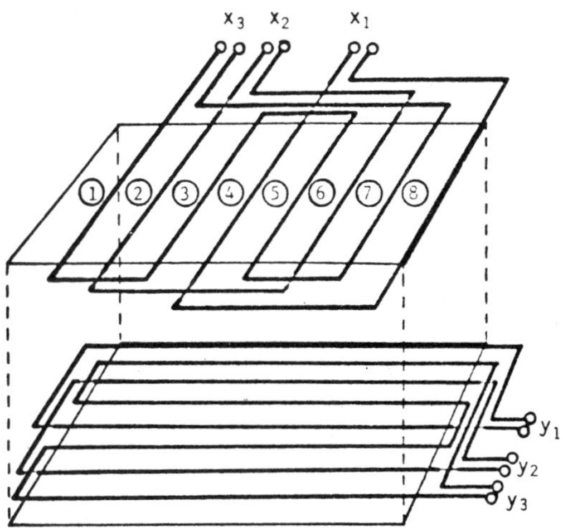

Fig. 2. Gray-coded detecting plane

Region	X_{n-4}	X_{n-3}	X_{n-2}	X_{n-1}	X_n
①	0	0	0	0	0
②	0	0	0	0	1
③	0	0	0	1	1
④	0	0	1	1	1
⑤	0	1	1	1	1
⑥	0	1	1	1	0
⑦	0	1	1	0	0
⑧	0	1	0	0	0
⑨	1	1	0	0	0
⑩	1	1	0	0	1
⑪	1	1	0	1	1
⑫	1	1	1	1	1
⑬	1	0	1	1	1
⑭	1	0	1	1	0
⑮	1	0	1	0	0
⑯	1	0	0	0	0

Table 2. Alternative unit distance code

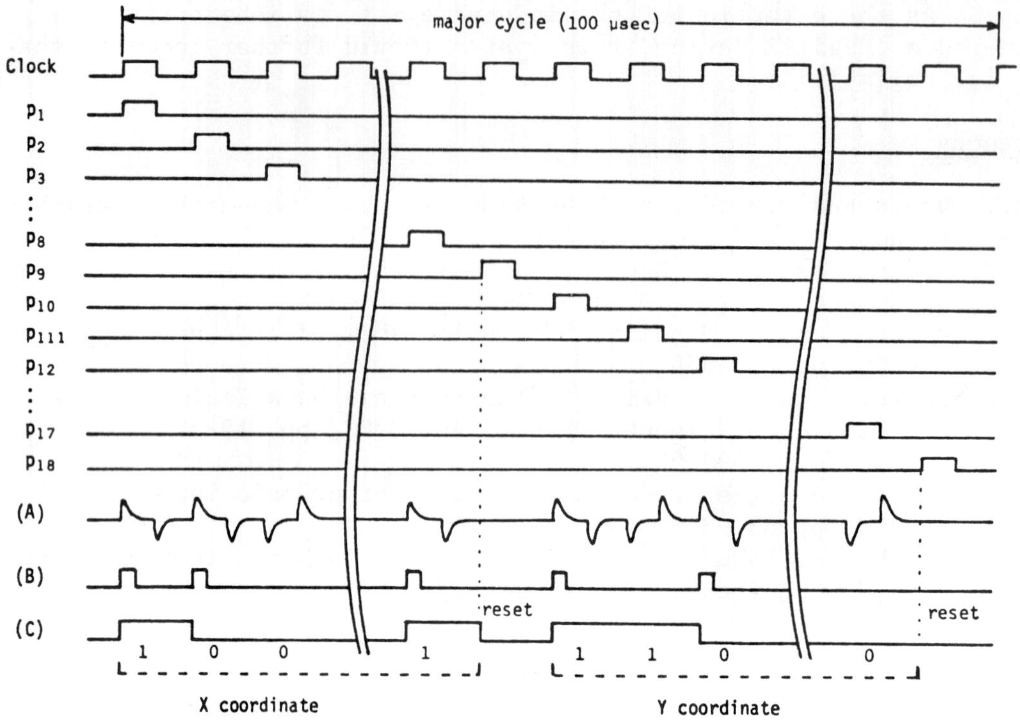

Fig. 3. Timing diagram

Fig. 4. Block diagram

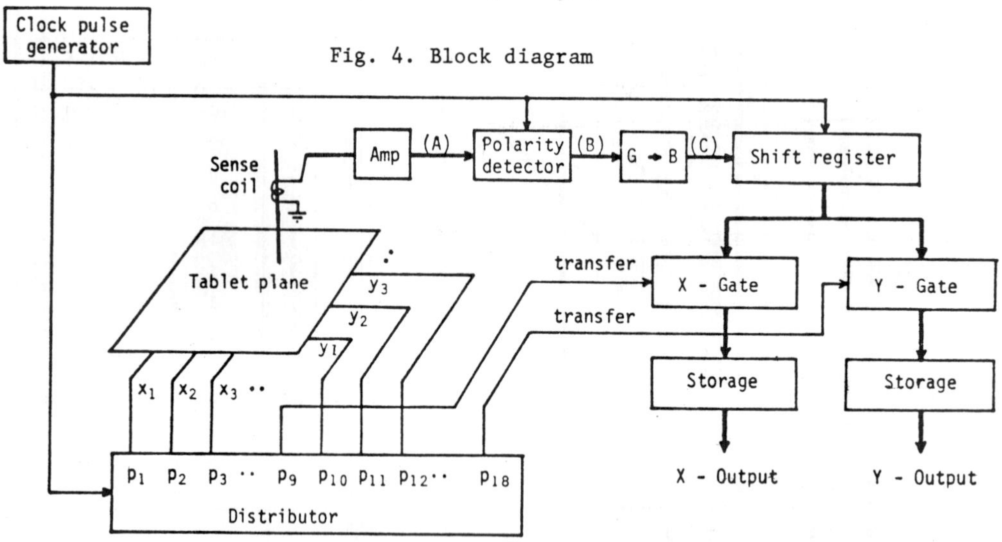

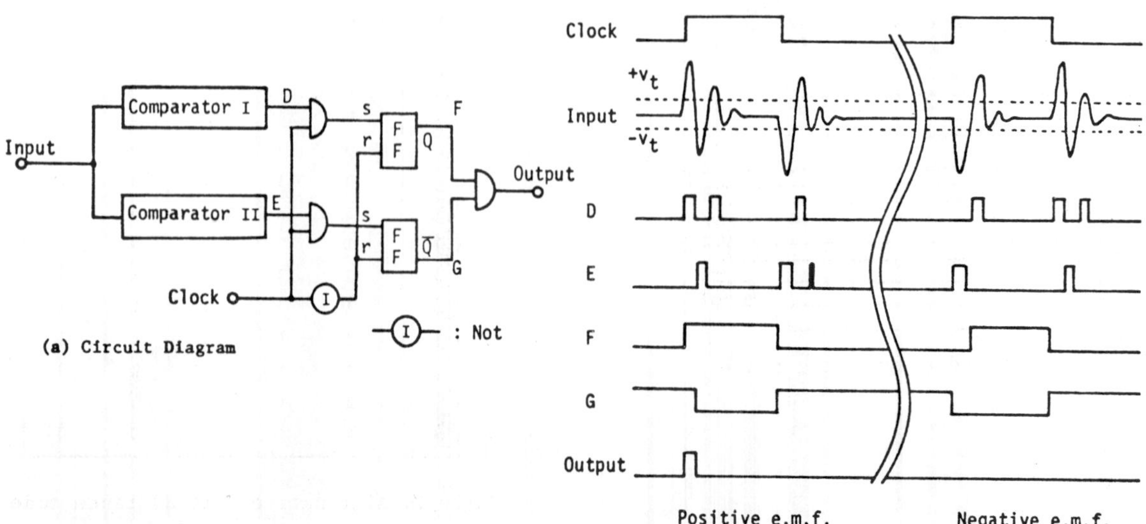

(a) Circuit Diagram

(b) Signal waveforms

Fig. 5. Polarity detecting circuit

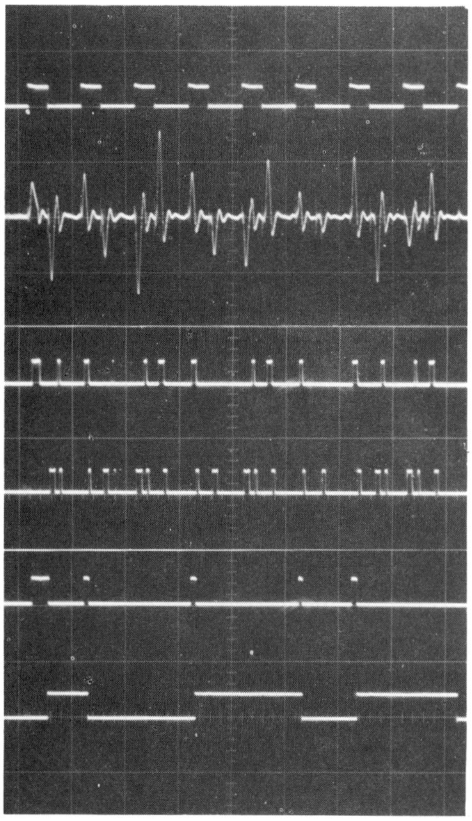

Fig. 6. Oscilloscope patterns
Clock pulse, Induced e.m.f., Output of Comparator I, Output of Comparator II, Gray code signal, Binary output.

Fig.7. Prototype model and monitored pattern.

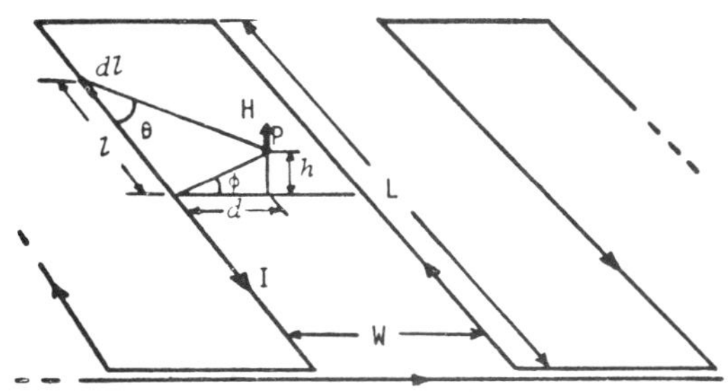

(a) Tablet plane

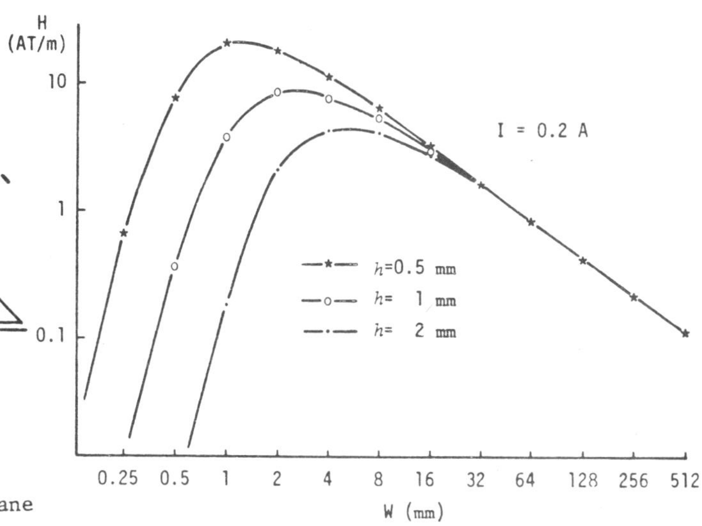

Fig. 8. Magnetic field on tablet plane

(b) Magnetic field intensity

AREA MEASUREMENT USING TELEVISION AND DIGITAL TECHNIQUES

D J Pilkington

SUMMARY

This paper outlines the development of a new instrument for measuring the areas of flat shapes, such as leather skins or shoe patterns.

The waveform generated in a slightly modified closed-circuit television camera is processed in digital circuitry enabling the time spent scanning the object, which is proportional to the required area, to be measured.

By suitable calibration techniques the instrument can directly read out area in any units, the time for measurement being less than a second. Physically, the object should be in the field of view of the camera and in high optical contrast with its background.

INTRODUCTION

In the process of leather manufacture the skins are subjected to drying operations which necessitate them being stretched on frames before the drying period. Fluctuations in area take place during drying and subsequent operations, and a study of these changes in area was requested by a SATRA member, hopefully to establish whether area was being lost. Since leather is sold by area there is an incentive to reduce losses in area that occur because of the production process.

The skins in leather manufacture are highly irregular in shape and can be up to 2.6 m^2 in area. To measure the area when stretched out on drying frames is not possible with existing leather area measuring equipment. Consideration was given to photographic techniques, where the profile of the skin would be cut out from the print and weighed, the weight being proportional to area. This technique was not considered accurate enough when three pieces of paper of equal size had 2 % weight variations.

In shoemaking, the shoemaker checks the area of skins supplied by the leather manufacturer by counting the number of squares of known area when a Perspex graticule is placed on the skin. This technique was too slow for use in the tannery investigations.

The specification for the required measuring system could be summarised thus :-

1. Portable; to be operable in various positions in the tannery.
2. High resolution; able to detect changes in area of 1 %.
3. Remote; to have no physical contact with or close proximity to the skin.
4. Rapid in operation; minimal interference with normal production activities.

D J Pilkington is with The Shoe & Allied Trades Research Association, Kettering

PRINCIPLES OF THE TELEVISION SYSTEM

To comply with the above specification it was considered that a television camera was a convenient way of scanning a skin, provided the signal generated could be easily converted to a number equal or proportional to area. Figure 1 illustrates the principle of operation.

When a television camera registers the image of an object it divides the field of view into a number of equally spaced horizontal lines, such as AB and CD in Figure 1. The scanning velocity is assumed to be constant and, therefore, the time required to scan one line is proportional to the distance traversed during the scan. Totalising the distance traversed during the scanning of a skin gives a figure which is proportional to the skin area. To measure an area, a facility is required to extract from the camera's output waveform only the time spent scanning the skin. By comparing this time with the time required to scan a known calibration area, the unknown area can be computed.

A physical requirement is that the skin, or object to be measured, must be in high optical contrast with its background. It has been found with the use of the equipment that back lighting systems are the most effective.

To measure the time spent scanning the skin a standard electronic digital counter is adequate. However, an interface unit between the camera and the counter is necessary to ensure that the measurement takes place during an integral number of frames. It is also useful to be able to initiate the measurement by pushing a button or some form of automatic starter.

Figure 1

Principle of Television Area Measurement

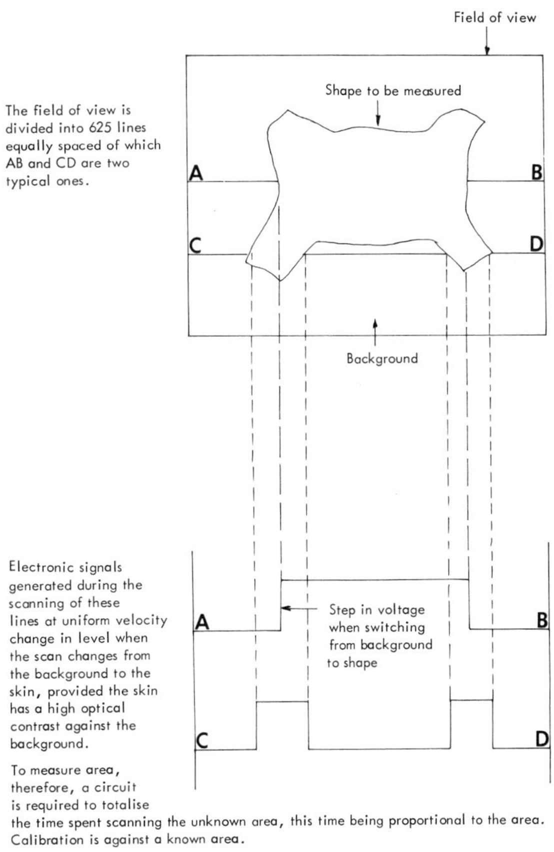

SATRA developed such an interface unit and it is shown in block diagram form in Figure 2. The waveform emanating from the camera is a composite one containing video signals, which have the different levels of voltage for the skin and its background, and line and field synchronising pulses. This is illustrated in Figure 3.

The video separater, block 3 in Figure 2, extracts the video content of the composite waveform, using standard differential amplifier techniques. A similar type of amplifier is used to extract the synchronising pulses used in the field pulse generator, block 1 in Figure 2. The line pulses are filtered out leaving a train of field pulses, the period of which is 20ms Dividing this train by four produces a square wave form of 40ms high period, followed by 40ms low period. 40ms is, of course, the time for one complete frame to be scanned. When more than one frame is to be scanned then additional division stages will produce longer pulses.

This waveform, equal in length to the time for one complete frame or multiple of frames, is linked with the synchronising circuit, block 2 in Figure 2, which contains the push button to start the measurement. This ensures that the measurement does not commence until the start of the first frame after the button has been pushed. In block 4 the video waveform and the synchronising waveform are gated together to produce a waveform containing the video content for one or more frames. It is usually arranged that the higher voltage level in this waveform corresponds to the scanning of the object and the lower voltage level to the scanning of the background.

The output waveform from block 4 is gated with a 10 MHz crystal controlled square wave oscillator, the subsequent train of pulses being fed into a variable divider unit, block 5. This variable division circuit uses four decade high speed pre-set counters and controls the number of pulses to be counted in an orthodox up-counter and display section, block 7 in Figure 2.

MEASURING PROCEDURE

The camera is set in a position such that the object to be measured fills as much of the field of view as possible. When this position is established the known calibra-

Figure 2

Area Measurement (Block diagram)

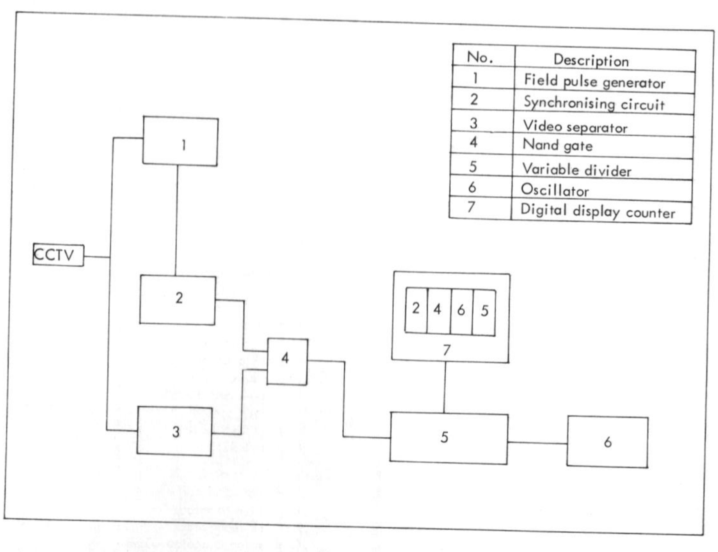

tion area, which is usually a rectangle, is set up for measurement. The calibration area is programmed in on the thumb wheel switches of the variable divider unit, the area being in any selected units. On taking a measurement the read-out obtained is then programmed on the thumb wheel switches. With this new division factor, the area measured when the button is pushed again will be displayed as the calibration area. Providing the aperture setting of the camera and its distance from the plane of measurement is not altered, the equipment is then ready for measuring unknown areas. It takes considerably less than one second to measure the area of the object once it is in position.

Checking and setting up as just described takes only a few minutes and changes from one unit of area to another are quickly achieved by simple re-calibration.

APPLICATIONS

Leather manufacture : The equipment has been used in tanneries to monitor the changes in the area of skins during drying and finishing operations, and over 10,000 readings have been taken.

Shoe manufacture : The patterns, which form the components that constitute the uppers of shoes, are irregularly shaped flat pieces and, therefore, amenable to measurement by the technique described. When a new design of shoe is produced an estimate of the area of leather or other upper material required to produce the batch has to be made. To provide this estimate, absolute pattern areas and interlock areas are required. The interlock area allows for inevitable wastage

Figure 3 Simplified wave form from camera

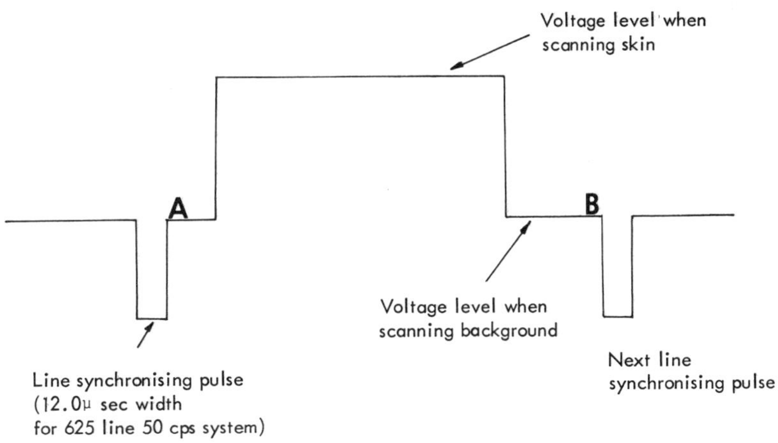

After ½ frame, or 1 field, which corresponds to 313 lines, a field synchronising pulse is generated of a greater width than line synchronising pulses

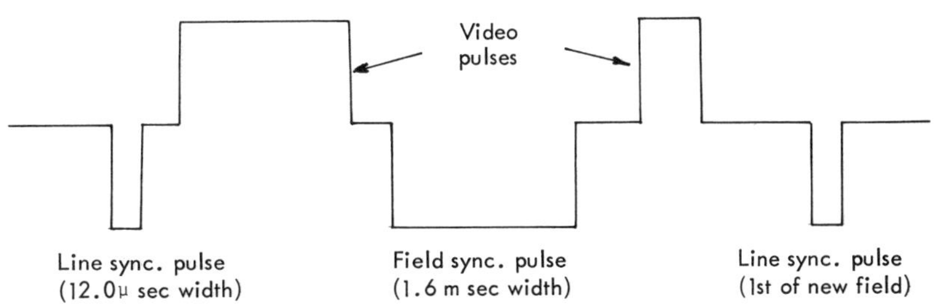

between adjacent patterns when laid on the skin from which they are being produced. Traditionally interlock areas are measured using reasonably complex geometrical constructions and planimeters are used to measure absolute pattern areas.

The television digital technique is capable of both types of measurement and, because of its speed of operation, it is estimated that up to 75 % time saving will be made when completing the leather estimates for a range of patterns.

Table 1 shows 12 successive readings taken for two areas of 50 and 25 dm^2 but with the same camera distance and calibration setting. A clear indication of repeatability and accuracy can be deduced. Figure 4 shows a photograph of the equipment measuring a skin stretched on a drying frame.

Other applications : The early work with the equipment naturally concentrated on its use in the shoe industry and the supply trade to that industry, but SATRA is not excluded from supplying the equipment to be used in other types of measurement. With the equipment having a remote scanner it is possible to measure small as well as large areas, to monitor changes in area through process, and to look at sections of 3-dimensional objects.

By using TTL integrated circuits throughout, the equipment has proved to be very reliable and approaching 20,000 readings have been taken with the prototype equipment. Being digital in nature, it is ideally suited to printing out areas and recording data on punched tape.

TABLE 1

Test Area	50 dm^2.	25 dm^2.
Calibration Settings	227	227
Area Read outs	50.02	25.02
	50.04	25.06
	50.06	25.05
	50.01	25.07
	50.07	25.05
	50.01	25.05
	50.03	25.05
	50.02	25.03
	50.06	25.03
	50.02	25.04
	49.97	25.03
	50.03	25.02
Mean	50.03	25.04
Range of Readings	0.1	0.05
Standard Deviation	0.03	0.02
or	0:06 %	0.08 %

CONCLUSIONS

There follows a summary of the advantages of this type of measuring equipment :-

1. Remote operation
This enables large components as well as small to be measured without interference by proximity sensors or contact devices.

2. High resolution
The use of 625 line television systems and 10 MHz pulse frequencies means that the vertical resolution is, at best, one part in 575 and the horizontal resolution is, at best, one part in 520.

3. Standard camera equipment
The type of camera used is a general purpose close circuit television camera, designed for application in rugged invironments. Some minor but vital alterations are necessary to produce the required waveform, so it is ill-advised to consider that the digital equipment is automatically suited to any camera.

4. Competitive price
Because so much standard equipment is used, the cost of the basic equipment which comprises a camera, monitor and digital unit, will compare favourably with existing measuring systems for areas of irregular shape.

ACKNOWLEDGEMENTS

The author wishes to acknowledge the assistance given by his colleagues at SATRA for their work in the development and application of the area measuring equipment.

Also the author would like to thank Dr A R Payne, Director of SATRA, for his permission to publish this paper.

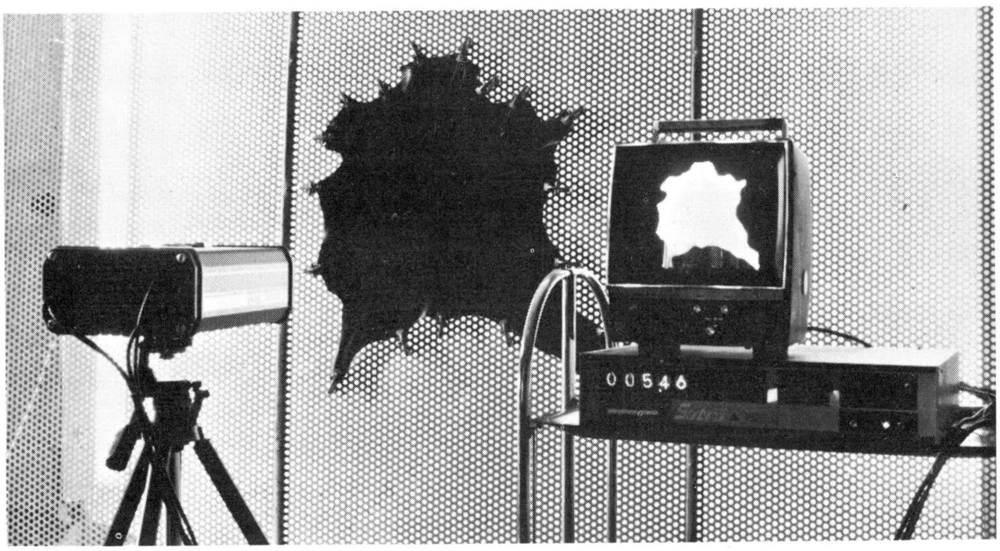

Figure 4

AN AUTOMATIC FRINGE SCANNER FOR STRAIN MEASUREMENT

D. J. Davies*, W. A. Evans* and A. Luxmoore**

1.0 INTRODUCTION: It is well known phenomenon that when two periodic or near periodic structures are overlapped and viewed in appropriate lighting conditions, patterns of light and dark fringes are formed. The effect was first observed in "watered" silk from which the name moire is derived (1). Everyday examples include certain modern semi-transparent fabrics and television pictures where the line scan may interact with a periodic structure being televised. The earliest reference to scientific use of moire fringes is due to Lord Rayleigh (2) in 1874, who suggested that periodic errors in ruled plates could easily be detected by placing two such plates together with a slight angle between the two line directions. Fringes would be produced and irregularities in the fringes could signify irregularities in the rulings of the gratings. Actual use of moire fringes for scientific measurement was to come much later.

The use of moire fringes for strain measurement has been well documented by Hitchings (3). The underlying principle is based on planting on the test member a grid by an etching or similar process. The grid is viewed through an identical grid in close proximity whilst the test member is stressed. The change in line density of the etched grid causes moire fringes to appear as seen through the reference grid. Information on the degree of compression or stretching of the specimen is contained in the distances between the peaks and troughs of these fringes. Any irregularities in the moire fringes produced serve as an indication of local stress variations, and this feature, in particular, has proved beneficial to workers using the method. The usual technique of information retrieval is via enlarged images of fringes projected on to a screen from a photographic negative with the required distances being measured with a ruler. An accuracy of the order of 10% of the distances between peaks of intensity is claimed for the method.

Attempts at automation of the measurements usually have in common the use of a photographic record. Theocaris (4) uses a technique of scanning a photographic record past an aperture behind which is placed a system for measurement of line intensity. The voltage output is recorded on paper tape, where the drive of the paper tape deck is in correspondance with the movement of the photographic plate across the aperture. The overall effect is an extension of the horizontal axis many times from the original, from which measurement of the distances between peaks and troughs of the moire fringes can be determined with improved accuracy. Sciamarella (5) uses a similar system except that the voltage output of the densitometer is converted to digital information and stored on paper tape for further processing. Hitchings (3) chose a single mirror scanning system for his work, where the image was focussed on a photosensor to yield a measure of the light intensity distribution of the fringes with time, and hence with linear distance, since the distance covered by the scan is proportional to the time of the scan. The processor was based on digital counting from a gated master oscillator and included a core store for storage of the data. The

* Department of Electrical and Electronic Engineering
** Department of Civil Engineering
University College of Wales, Swansea, SA2 8PP.

system could be considered as some advance on previous methods in that the time consuming photographic step was no longer necessary, but multiline scanning, so essential to the user, did not figure in the instrument. Besides, several engineering features required further development in the early prototype.

In the light of previous work it was felt necessary to totally reappraise the problem with the view to eliminating the obvious weaknesses of previous efforts at automation. A scanning system based on an all electronic approach was adopted, and whilst it is not claimed that all the problems have been solved, the instrument described below offers a flexibility to the user not contemplated in earlier designs.

2.0 INSTRUMENT DESCRIPTION: An overall view of the instrument system is shown in Fig. 4. A closed circuit television camera is used to scan the fringe pattern under observation and the result displayed on a monitor, an example of which appears in Fig. 2 with the intensified selected line position in the centre of the field of view. The intensified line may be moved up and down the field of view as line positions are selected and stored. This differential scan mode allows the user to concentrate on regions of particular interest. Alternatively, the user may select the number of lines required in the scan cycle up to a maximum of fifteen. The user is presented with a choice in processing options: the logic can either store the fringe positions, fringe separation or the average fringe separation of each line scanned. A line printer is incorporated as an output peripheral of the system. The output format is such that the results concerning a scan line are printed followed by the scan reference number, so that the line in question may be called up on the monitor if required. Absolute measurements with the instrument are relatively easy to arrange by the simple expedient of placing a reference plate in the focal plane of the camera once it is set up. Scanning the reference plate blanked with a square of accurately known dimensions introduces a normalising factor which, for convenience, is separately available on a seven segment display readout. The various sections of the instrument are shown in the block schematic of Fig. 1 and will be treated in as much detail as the space allows.

3.0 DISPLAY SYSTEM: The display unit incorporated in the instrument has no internal sweep circuitry, but contains high resolution x, y and z amplifiers each having differential inputs. This display unit was chosen because the means to display both single line and a complete video frame is easily arranged. In the display mode the camera sweep waveforms with appropriate interfacing are used to drive the x and y amplifiers. The video signal with interfacing is applied to the z input of the display monitor. Selected line intensification is under control of a bank of decade thumbwheel switches which form one input to a digital comparator. The second input is derived from a counter used to count the synchronising pulses of the composite video output signal of the camera. The comparator is used in the line intensification circuits whereby the display beam is switched to full intensity for the selected line of the frame giving a z modulation effect. For display of a single line output signal from the camera, a similar technique is employed with the proviso that the video signal is switched to the display y amplifier and the z amplifier input is grounded. Blanking must be incorporated excepting the selected line and no other.

The camera contains an electrostatic rather than an electromagnetic vidicon tube. Linearity of better than 1% is quoted for the vidicon tube deflection characteristic whilst the use of electrostatic deflection allows operation at an alternative lower frequency. This is not necessarily the case when electromagnetic deflection is employed when loading problems arise due to the scan coils losing their reactive impedance at lower frequencies. The camera is a commercial type with a 2:1 interlace in the composite video output. For each camera frame there is effectively 312 scan lines only. When displaying the output signal, the observed trace will obviously differ for consecutive frames. Within the resolution of the camera this has been found quite acceptable without making further provision. Fig. 3 is an example of a single line trace corresponding to the intensified line in Fig. 2. The lower trace of Fig. 3 is due to the flyback effect.

4.0 CENTRAL PROCESSOR: The central processor provides outputs for the x and y scans as well as data to the printer. The linearity of the internal scan waveforms of the camera itself are insufficient for computation purposes and for this reason are disabled when computation takes place and replaced with waveforms from the central processor. The x sweep waveform is synthesized using a ten bit counter and d to a converter with reset occurring on counter overflow. The feature of this approach to ramp waveform generation, which makes it particularly attractive in this application, is the availability of a digital representation of the x position of the scanning beam at any instant in time. The y position of the beam may be obtained from a second d to a converter operating from the y scan buffer. Scan lines are first stored in the memory during a writing sequence with the user selecting directly, or indirectly, where each line scan is located on the fringe pattern. Where the user selects the "differential" scan, the scan positions are chosen directly in accord with the thumbwheel switch settings. Alternatively, if the "equidistant" scan mode is selected the thumbwheel switch settings inform the system electronics of the number of scan lines required up to a maximum of fifteen. The processor then proceeds to calculate the required positions storing the results in the same memory locations as before. During the scan routine the output of the y scan is switched to the high voltage y amplifier and the stored scan positions sequentially read from the memory. Peak detection circuitry acting on the video input provides the control signal for storing the contents of the x scan counter at each peak and trough to yield information on fringe positions. With some additional logic, fringe separation may, alternatively, be stored.

In practice, the user often requires information on the average spacing of a single line scan. As an example, this routine has been chosen for more detailed treatment with the flowchart appearing in Fig. 5 and the associated logic circuitry in Fig. 6. The method of division employed in this scheme is a variation of the pulse rate technique. The divisor is loaded into down counter I and the divident into H. The clock to H and I is enabled until H reaches the all zeros state, thus sending an equivalent number of clock pulses to I as the original accumulator sum. I is arranged as a frequency dividing circuit; the number of "load" pulses being the quotient of the division. These pulses are counted by D and copied into the memory, overwriting part of the fringe separation data for the first scan line. The counters and shift registers are then cleared and the average fringe separation for the next scan line calculated.

In addition to the above a printer output routine is required together with sequence control logic governing the operation of the logic. The system is so designed that additional routines may be added as required.

5.0 CONSTRUCTION: Simulation studies were carried out to establish the integrity of the logic before the machine was built. Extensive use of the "wire wrap" principle has been made in the construction with the logic mounted on a separate board capable of holding 140, 16 pin dual inline integrated circuits. Interdigitated geometry is used for the power supply rails printed on the board to yield a high packing density with at least the semblance of a ground plane.

6.0 CONCLUSION: In operation the instrument has proved successful and filled a much sought after requirement in strain analysis. Experience has shown that further work is needed to enable the machine to handle noisy video signals and work is in hand to improve this aspect of the performance.

7.0 REFERENCES:

(1) Oster, G. and Nishijimay, Y. "Moire Patterns" Scientific American (1963) May edn.

(2) Lord Rayleigh "On the manufacture and theory of diffraction gratings" The London Edinburgh and Dublic Philosophical Magazine and Journal of Sciences (1874), 47, 30, 81-93.

(3) Hitchings, D. J. "Data Acquisition from Moire Fringes by Scanning Techniques (the observation of a systems design exercise) Ph.D. Thesis (1970) University of Wales (Swansea).

(4) Theocaris, P. S. "Moire fringes in Strain Analysis" (1969) Pergamon Press.

(5) Sciamarella, C. A. "Basic optical law in the interpretation of Moire patterns applied to the analysis of strain. Part 1, Experimental Mechanics (1965), May edition.

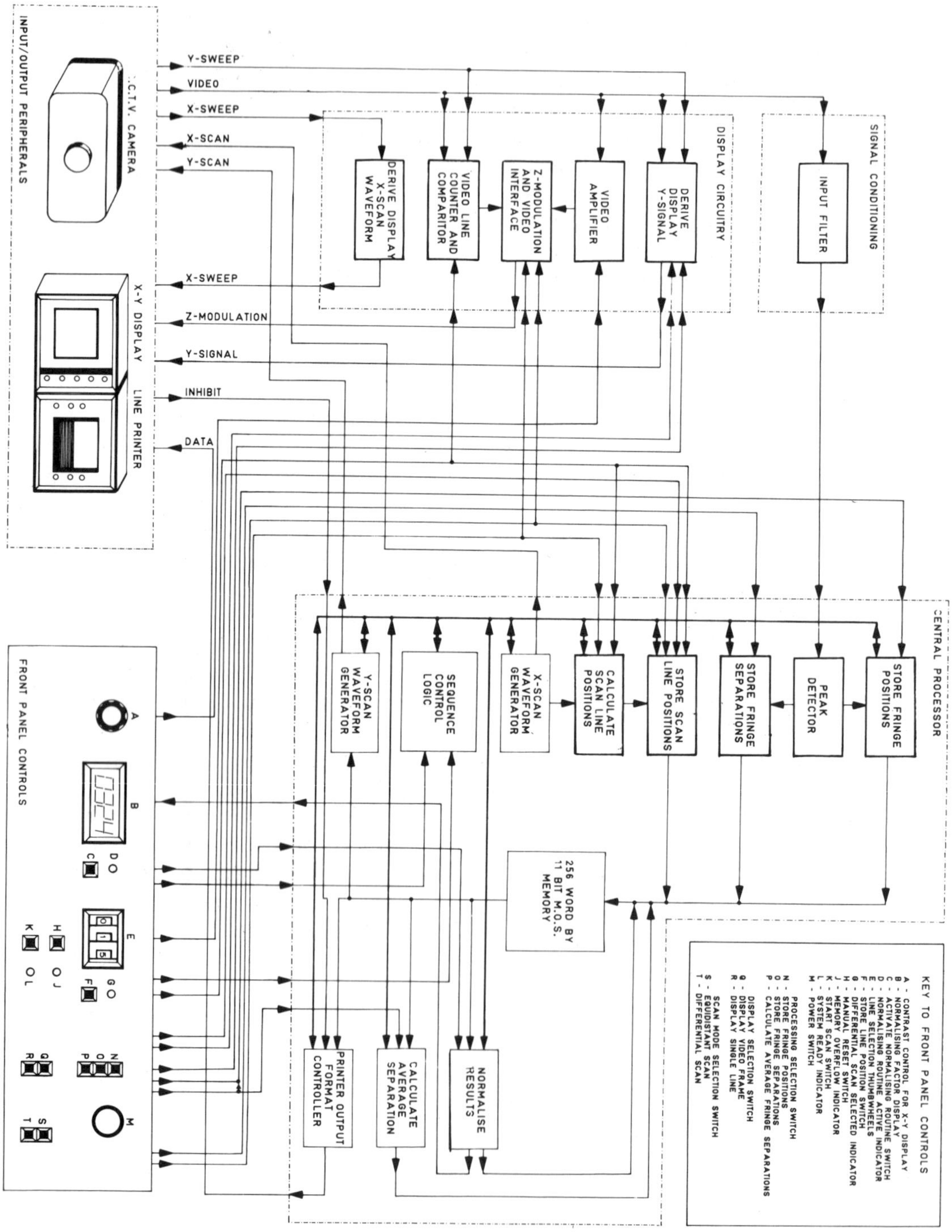

FIG. 1 BLOCK SCHEMATIC OF COMPLETE INSTRUMENT

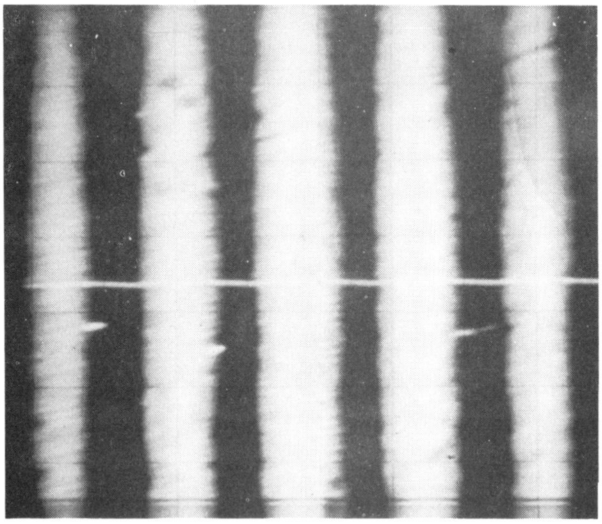

FIG. 2 MOIRÉ FRINGE PATTERN

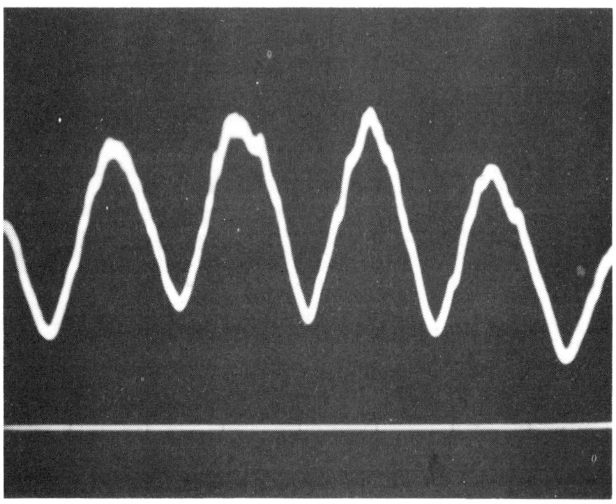

FIG. 3 SINGLE LINE TRACE

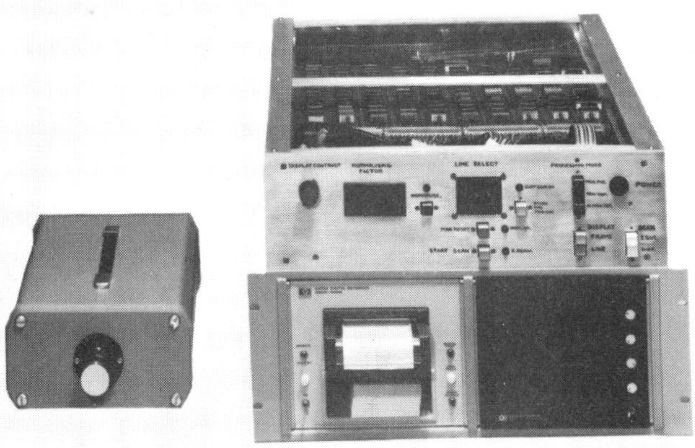

FIG. 4 VIEW OF INSTRUMENT

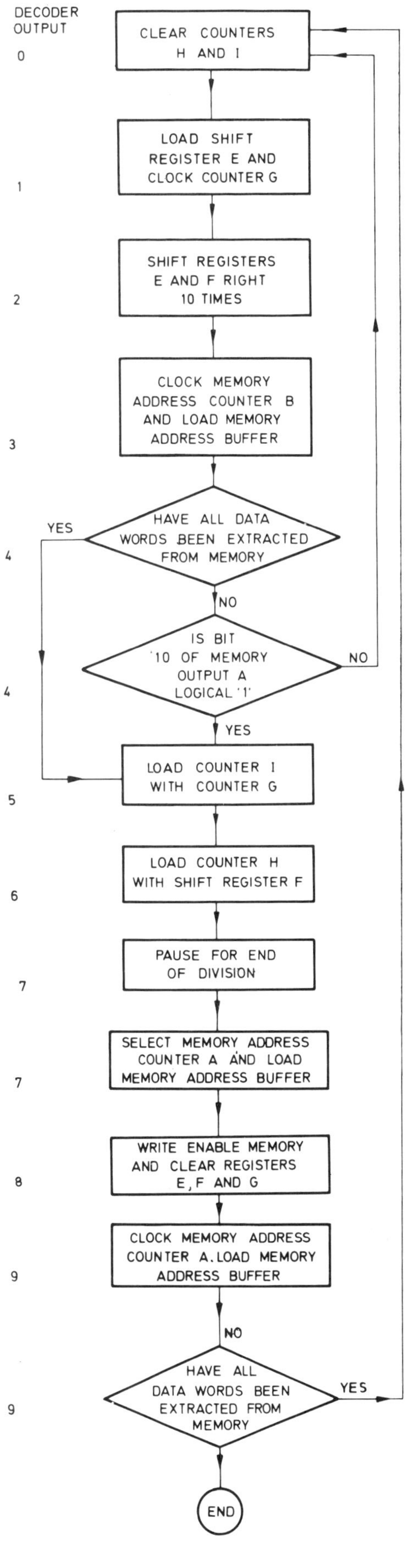

FIG. 5 FLOWCHART (AVERAGING ROUTINE)

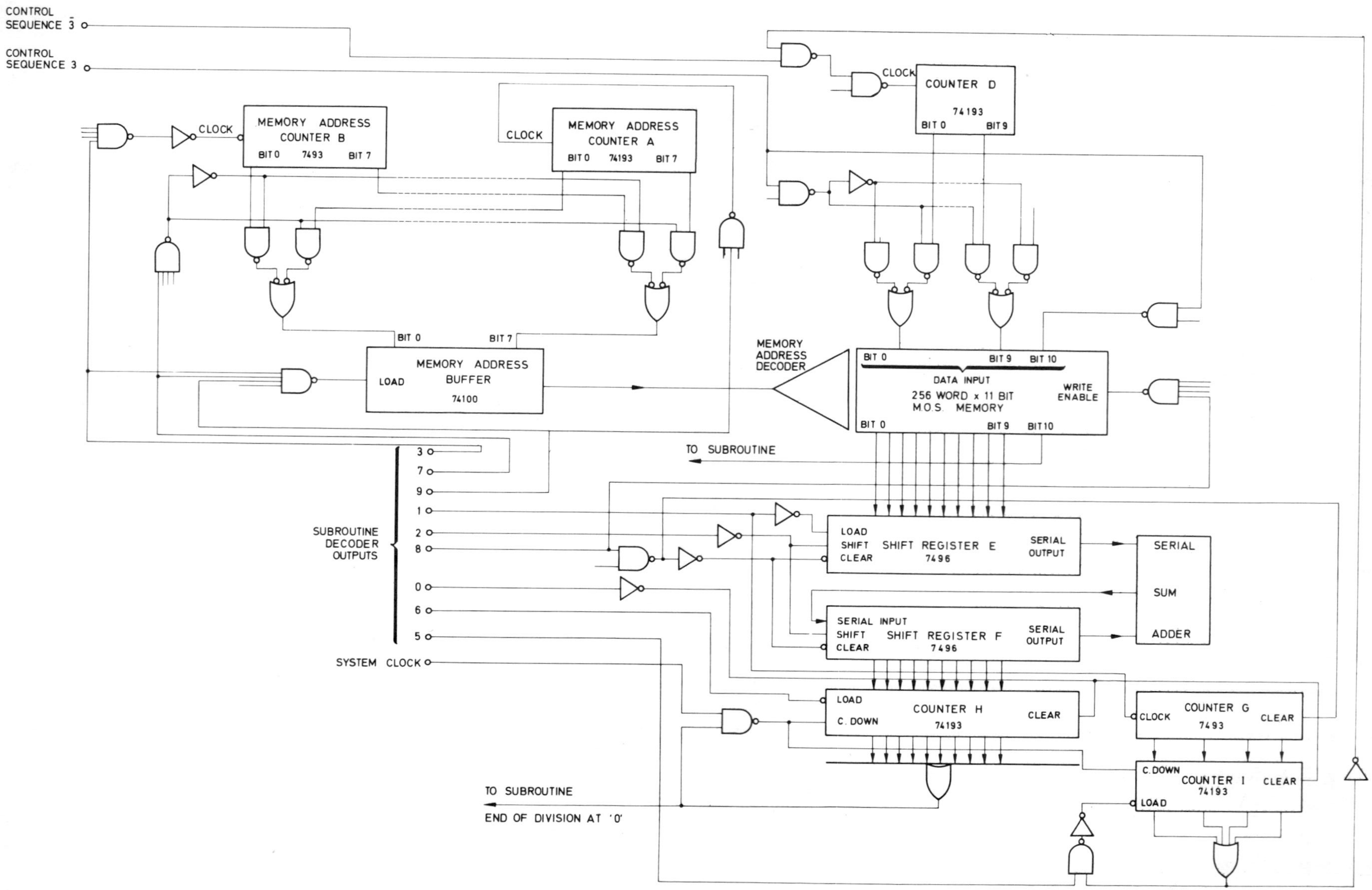

FIG. 6 EXAMPLE OF A ROUTINE (AVERAGING)

A NEW FORCE TO FREQUENCY TRANSDUCER

P.R. Wyman

1. Introduction. The frequency analogue of physical quantities is often a convenient one to use in digital instrumentation systems. The other choices available - voltage, current, parallel and serial coded highways - will no doubt be discussed elsewhere in this Conference, but a basic transducer element producing a frequency output determined by a physical input quantity should certainly be a useful component in systems using either frequency or digital code transmission. A number of such devices have appeared within the last decade or two. These include the Solartron vibrating tube densitometer (1), the Agar vibrating spool densitometer (1), the modulated radio-activity displacement transducers developed by Taylor and Rao at Strathclyde University, and the vibrating wire gauges used for measuring structural deflections. The latter devices bear some resemblance to the transducer which is the subject of this contribution, but there are several important differences. The vibrating-wire structural deflection gauge consists of a thin wire stretched between two points in a structure. The tension, and hence the natural frequency, of the wire is determined by factors including the mechanical strain and the Young's modulus. The wire is excited mechanically in its fundamental mode by plucking, and the frequency is measured electronically. Knowing the length of the wire and the density and Young's modulus of its material, the strain can be calculated for any observed frequency.

$$f_1 = \frac{1}{2L}\sqrt{\frac{S}{\rho}} \quad \text{(Reference 4)} \qquad \ldots(1)$$

f_1 = frequency of fundamental mode of transverse vibration
L = length of wire
S = axial stress
ρ = density

$$\text{Whence the extension} = \frac{4\rho f_1^2 L^3}{E} \qquad \ldots(2)$$

These quantities will usually be known well enough for this to be a useful method of surveying the deflections of loaded structures.

In process instrumentation systems, the physical variables which are most commonly transduced are pressure, differential pressure (often generated by flow through an orifice), and temperature. Temperatures are normally sensed electrically. Absolute, gauge, and differential pressures are measured by arranging for them to exert a force on a containing chamber. The pressure is inferred either by relating the deflection of the chamber to the elastic moduli of its material(s), or by generating a force in a linear "force motor" which is arranged to oppose the force due to the pressure, so that the deflection of the chamber is virtually zero. This latter method is used in the so-called "force balance"transmitters. Successful design according to the former method requires close control of the elastic moduli, and until comparatively recently the best process pressure trans-

Mr.P.R.Wyman is with The General Electric Co.Ltd., Hirst Research Centre, Wembley, Middlesex.

mitters all worked on force balance principles. Some very good micro-displacement transmitters have recently appeared - the motive for using this principle is the lower working power compared with that necessary to drive a force motor. This facilitates the use of the two-wire 4 to 20 mA analogue transmission system, particularly in explosive atmospheres.

Hence a force to frequency transducer which requires little driving power and has a low mechanical compliance should be an interesting device. The transducer which is the subject of this contribution combines these merits with a high degree of stability and mechanical simplicity. It was developed by The General Electric Company Limited at the Hirst Research Centre. It is, of course, tempting providence to describe, as is done in this contribution's title, any form of transducer as "new". Careful enquiries have not revealed any anticipation.

2. General Principles. The fundamental component of this transducer is a taut wire which interacts with a transverse magnetic field. Such a wire has an electrical impedance which peaks sharply at its frequencies of transverse resonance. It thus forms a two-terminal electrical network rather similar to a parallel L-C circuit, with the feature that its resonant frequencies can be modulated by the tension of the wire. Reference to any electronics text book will reveal numerous configurations of oscillator circuit whose frequency is set by the resonant frequency of a two-terminal tuned circuit, and in principle a taut wire with a transverse magnetic field could be used in any of these. In practice the low impedance level and the performance aims restrict the choice, and the circuit used is described later.

3. Resonant Member. Fig. 1 shows the main features of the resonant member. The tensioned wire is stretched between the poles of a permanent magnet. The magnetic flux density does not affect the resonant frequency of the wire; it only influences the impedance of the wire at this frequency.

Equation (1) applies to this system, and can be written:-

$$f_1 = \frac{1}{2L}\sqrt{\frac{T}{m}} \qquad \ldots(3)$$

where T = tension force (N)

m = mass per unit length (kg/m)

The impedance at the fundamental resonant frequency:

$$Z_1 = \frac{\pi Q B^2 L^2}{12\sqrt{mT}} \qquad \ldots(4)$$

where Q = quality factor

B = transverse flux density (T)

This impedance component adds to the wire's resistance and inductive reactance. Similar but smaller, components exist at the frequencies of the higher-mode resonances.

f_1 will be modified by the stiffness of the wire as a beam.

$$f_1' = f_1(1 + G) \qquad \ldots(5)$$

where $G = \dfrac{\pi^3 d^4 E}{128\, L^2 T}$ (d = wire diameter)

An example is given below of the combination of properties that can be achieved.

Titanium wire	0.063 mm diameter 32.0 mm long
Tension	0.81 N
Stress	2.62×10^8 N/m^2 ≈ 27% UTS
Resonant frequency	3766 Hz
Stiffness factor G	5.3×10^{-4}
Extension of wire under load	0.08 mm
Typical resonant impedance	2-3 ohm

Titanium is mechanically very suitable for the wire. It has the disadvantage of a high electrical resistivity, which in the example quoted above yields a wire resistance of 4.3 Ω, but this is not a problem because due allowance can be made in the oscillator circuit.

It is particularly important to terminate the wire efficiently, so that no relative movement exists between the wire and any part of the clamping arrangements. Furthermore no work hardening or embrittlement should occur at the terminations. Experiments made so far indicate that a satisfactory technique has been achieved.

It is essential in any application to ensure that the wire tension is limited to a safe value. Devices such as the closed-coil spring offer convenient ways of achieving this.

The only apparent restriction to increasing the sensitivity is the thinness of the wire that would be required. With care it should be possible to make a device that will resolve a few tens of micronewtons. In the other direction, acceptable linearity dictates a wire length which increases with the rating and is in the region of 1 metre at 10 kN (∿1 ton.f). Though such a device could be made, its bulk would make it less acceptable than the competing strain gauge load cell. The practicable limit may be around 100N.

4. Oscillator Circuit. Fig.2 illustrates an oscillator circuit embodying a vibrating wire element. The principle is that of an amplifier with broadband negative feedback, and narrow band positive feedback. The output amplitude is regulated by controlling the loop gain with the field effect transistor. Two 741 operational amplifiers in cascade are used to provide adequate gain so that operation is maintained over a wide frequency range. The oscillator circuit can be adapted to offset the component of feedback due to voltage drop in the wire resistance, although this is often not necessary.

5. Frequency Squaring. It will be apparent from Equation (3) that the output frequency is proportional to the square root of the tension force. When the tension force is derived from the differential pressure developed by a venturi or orifice plate, this has the happy result that the frequency is proportional to the flow rate. In general, however, a means of producing an output signal proportional to the square of the wire frequency will be required. This output signal can be a frequency, a pulse interval, or indeed it can be a current. Methods of producing these are discussed below (3).

Probably the simplest method of producing an output signal whose mean frequency is proportional to the square of the source frequency (f_1) is one using binary rate multipliers (BRMs), such as is illustrated in Fig.3(a).

In all these conversion circuits a frequency reference is necessary which will normally be stabilised by a quartz crystal, and this is denoted in this figure as f_x. The storage register holds a number equal to $\frac{nf_1}{f_x}$, which is applied to the control inputs of the BRM. The output of the BRM consists of a train of pulses of mean frequency $f_1^2 \times \frac{n}{m\, f_x}$, where m is a constant determined by the BRM configuration.

A disadvantage of this circuit is the fact that the output pulses are irregular, and one way of avoiding this is shown in Fig.3(b). Here the BRM is replaced by a phase locked loop in which a frequency (f_o) is generated by a voltage controlled oscillator, divided by a number proportional to f_1, and the result $\frac{f_o}{nf_1}$ compared phase-wise with f_1. The V.C.O. frequency is servoed so that $\frac{n_o}{nf_1} = f_1$, thus $f_o = f_1^2 \times n$. It may then be necessary to divide f_o by a fixed ratio to get a conveniently transmissible frequency.

Fig.3(c) illustrates a method for producing a regular pulse train output signal whose period is proportional to the square of the wire frequency. Using BRMs and counters, a number is set up in the storage register proportional to f_1^2. In another counter, the reference frequency is divided by this number. It is possible more simply to generate an irregular pulse train whose mean period is proportional to the square of the wire frequency, but the main advantage of period-proportional signalling over frequency-proportional signalling is that a linear reading of high resolution can be made in a receiver more quickly. This advantage is lost if the pulses are irregularly spaced.

Operational amplifier methods can be developed for generating an output current proportional to the square of the wire frequency. If the very low power amplifiers now available are used, this should enable the high precision features of this transducer to be exploited in two-wire 4 to 20 mA instrumentation.

6. Performance.

Temperature Effects

Theoretically one would expect a temperature coefficient of frequency equal to minus half the expansion coefficient of the wire material. Even when the frequency is squared, this is very small ($\sim$10 ppm per degree C for titanium) compared with other typical transducer coefficients (often upwards of 200 ppm per degree C). In practice performance is not quite as good as this, and 20-30 ppm per degree C for frequency squared is observed. The source of the discrepancy is not yet understood, but may be connected with temperature effects on the metallurgical hysteresis. The ultimate working temperature range of the transducer is not known, but must be very wide.

Linearity

The linearity of f^2 is also good. Such departure from linearity as does occur is caused: (a) by the extension of the wire under load, and (b) by the influence of the "stiffness factor" G in Equation (5). These combine to produce a typical terminal non-linearity of 0.2%.

Hysteresis

This is very small, typically 0.02% referred to f^2.

Repeatability (short term)
Experiments suggest that this is in the order of 0.05% (referred to f^2), but may have told us more about the loading system than about the transducer.

Repeatability (long term)
The standard deviation of twice-daily readings of a prototype taken over a period of 4 months was 0.01%, referred to f^2, after allowing for temperature fluctuation.

Response Time
The oscillator output frequency responds very rapidly to a step change in tension force. However, the frequency squaring systems described will in general respond much more slowly. If rapid modulation of the wire frequency exists, due, for example, to vibration along the wire axis, the squared output signal may be unreliable.

7. Conclusions. The two most obvious areas where applications may arise for this transducer are process and avionic instrumentation.

In process instrumentation the stability and precision compare favourably with alternative techniques, as does the linearity of the frequency squared output. In systems using digital or frequency-analogue transmission, where a square-root conversion law is preferred or acceptable, the vibrating wire transducer appears to offer significant advantages. If it is necessary to convert the output locally to a uniformly spaced pulse train whose period or frequency is proportional to the square of the wire frequency, the advantages over transducers generating linear voltage or current analogues may be less marked. Conversion to such modes from linear voltage is just about as easy as from square-root frequency. The fact that the transducer cannot work immersed in a liquid is a disadvantage discouraging its use in differential pressure transmitters. However, it may be that its precision and mechanical simplicity will be shown to outweigh these shortcomings.

In avionic instrumentation neither the square-root conversion law nor the inability to work under liquid is so significant as in process instrumentation. Here the stability, compactness, low mechanical compliance, and low temperature coefficient should be significant advantages.

Weighing systems of up to 100 N rating may be worth investigation.

References

(1) Wightman, E.J. Instrumentation in Process Control. Butterworths, London, 1972.
(2) British Patent Application No. 13549/73.
(3) British Patent Application No. 33907/73.
(4) Starling and Woodall, "Physics", Longman Queen and Co., London, 1957, 2nd edition, p. 847.

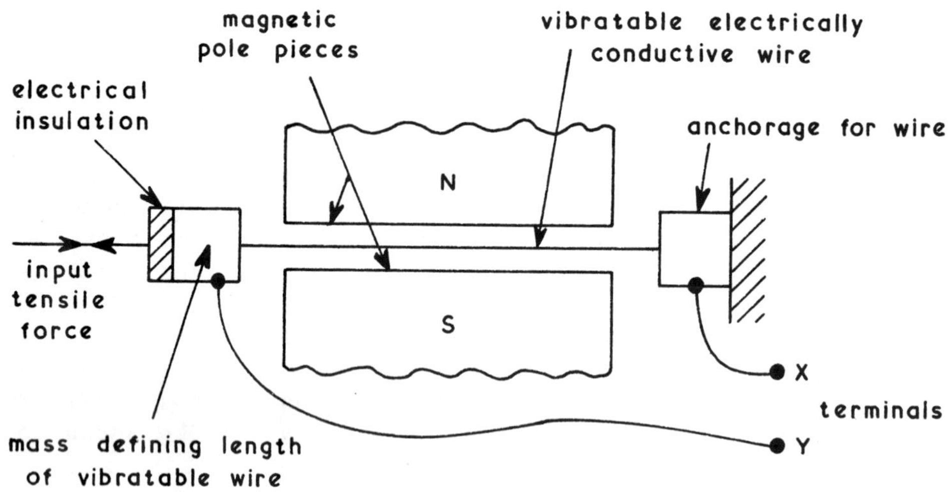

FIG. I. VIBRATING WIRE RESONANT CIRCUIT ELEMENT

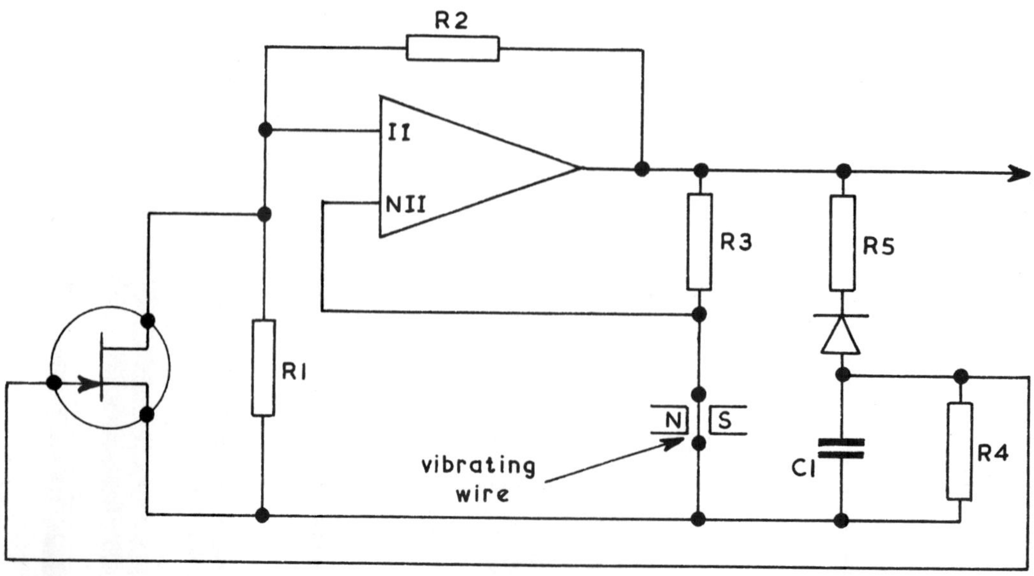

FIG. 2.
OSCILLATOR CIRCUIT TUNED BY VIBRATING WIRE RESONANT ELEMENT

FIG. 3.

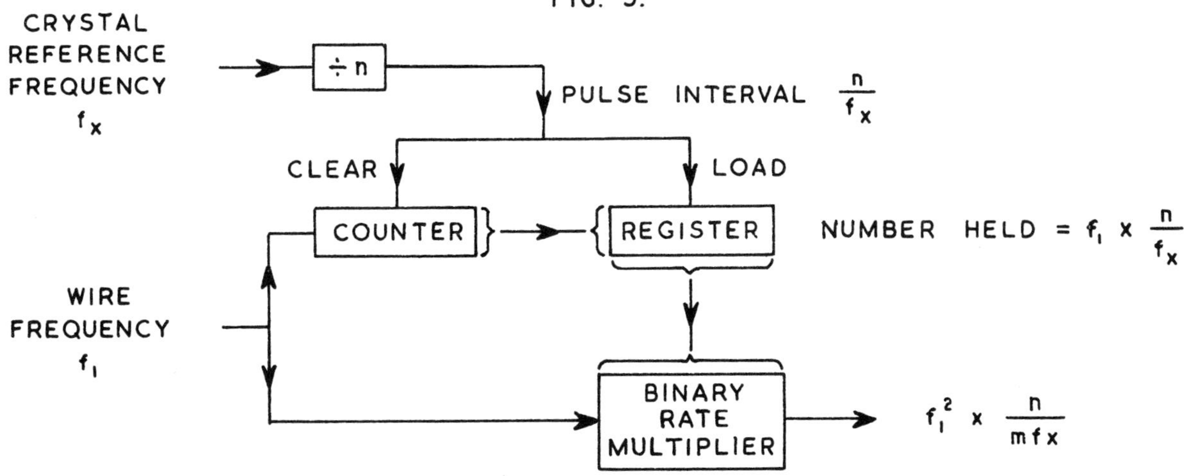

a. SIMPLE BRM FREQUENCY SQUARER.

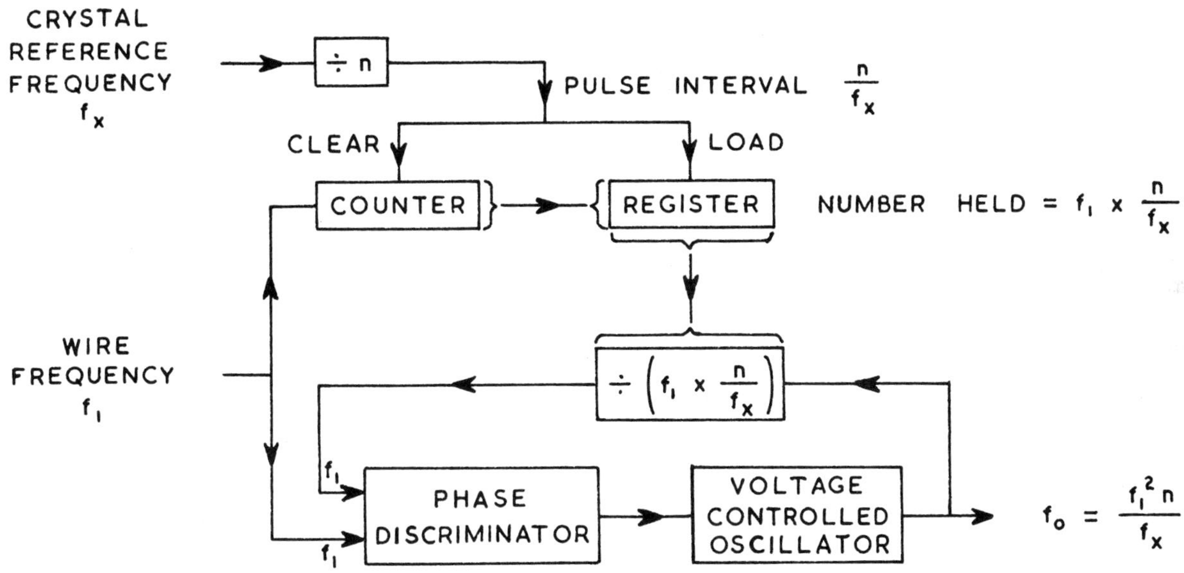

b. FREQUENCY SQUARER WITH REGULAR OUTPUT PULSES.

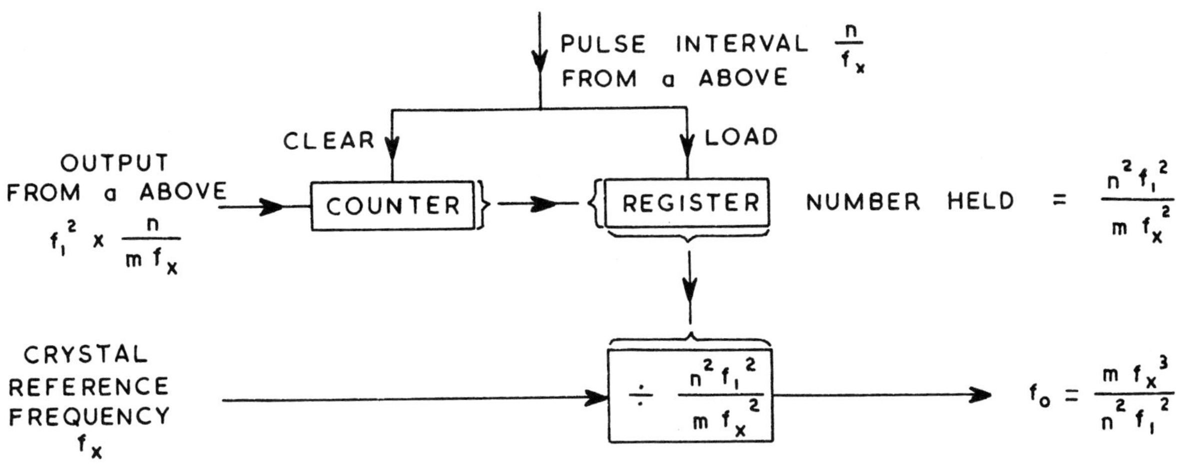

c. CONVERTER WITH OUTPUT PERIOD PROPORTIONAL TO SQUARE OF INPUT FREQUENCY.

A DIGITAL SIGNAL AVERAGER USING A GENERAL PURPOSE CONTROL MODULE

G. Erlandsson, G. Floom, B. Hallgren

1. Introduction

A digital averager system has been developed for measurements in microwave spectroscopy. The instrument is used for averaging repetitive noisy signals. Signal averaging has the ability to recover the signal's original waveshape from the noise. The signal-to-noise ratio can in this way be improved by several orders of magnitude. The instrument can be used in many fields of applications e.g. physical and biological sciences.

A measurement using the instrument is performed in the following way: The input signal is observed on an oscilloscope. The time-axis is divided into 256 points. For each point the signal is sampled and converted to digital form. The new value is added to the value accumulated in the memory. To start a sweep an external synchronizing signal must be available. The process continues for a preset number of sweeps and then the result in the memory is displayed on the oscilloscope.

After n sweeps the signal-to-noise ratio is improved by a factor n. Apart from random noise, other signals which are not time-locked to the sync pulse are also attenuated relative the desired signal.

2.1 Processor Structure

The instrument is organized as a 16 bit fixed program processor with a 256 word memory. It consists of an arithmetic unit (ARU) including two registers, a D/A-converter (DAC), an analogue comparator (COMP) and a Track-Hold unit (T/H). A simplified block diagram of the instrument is shown in Fig. 1.

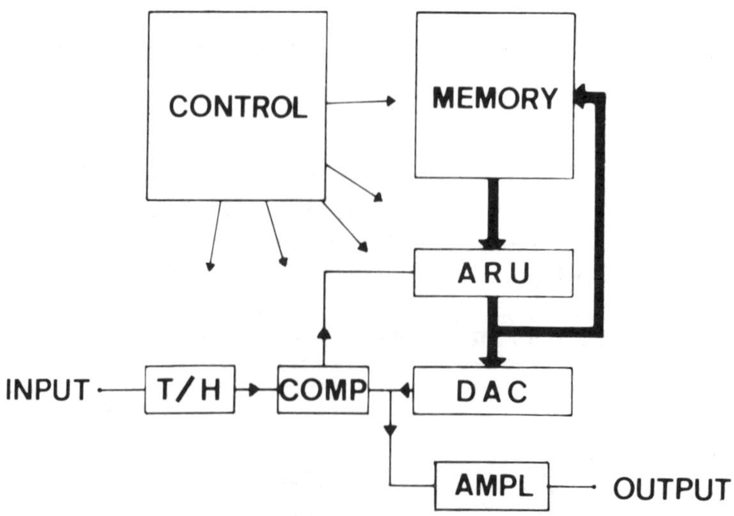

Fig. 1 Correlator Block Diagram

The authors are with the Department of Physics, University of Stockholm, S-113 46 Stockholm, Sweden

2.2 The Memory

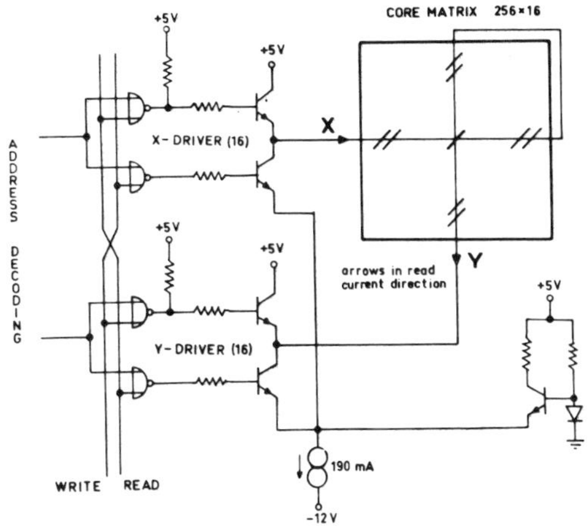

Fig. 2 Memory Drive System

The memory is built around a commercial core matrix consisting of 4 planes with 4x16x16 50 mil cores each. It is organized as a 3D coincident current system. The three drive currents 'X', 'Y', and 'Z' are reduced by a factor of two since each of the three wires is wired twice through each core. All 'X'- and 'Y' - wires are connected to each other at one end and the other end of each wire is connected to the midpoint of two transistors in a totem pole configuration. See Fig. 2. The emitters of the bottom transistors are connected to a common constant current generator of 190 mA. The base of each transistor is connected to a TTL-gate. By selecting the appropriate two transistors the selected word can be read or written. When no drive current is selected a separate transistor is conducting so that fast switching times can be achieved. The drive system has proved to give sense line signals with very little time and amplitude jitter. As sense amplifiers the linear IC 711 has been used.

The memory is operated in a split cycle mode. The read operation is started at the same time as the A/D-conversion procedure. The memory waits for the arithmetic operations to be completed and then the write operation begins. The memory cycle time is 5 us and the access time is 0.75 us.

When the memory was designed in 1969 alternative memory types were considered. A MOS-shiftregister was then rejected for cost reasons. Presently (1973) a memory constructed with MOS-shiftregister would cost about 1/10 of the core memory used with 10 times the speed.

2.3 The Arithmetic Unit

The arithmetic unit is built in 16 bit. It consists of a parallell load right shift register (T-register), a ripple carry adder and a parallell load accumulator register (A-register).

The arithmetic unit can perform three data operations: addition A+T, subtraction A-T and right shift of T-register T/2. Negative numbers are represented in 2's complement notation. To subtract the T-register from the A-register the contents of the T-register are complemented and added to the contents of the A-register with the adder carry-in equal to '1'. The total add or subtract time is about 120 ns for 16 bit.

2.4 The D/A-converter

The converter is time-shared in two applications: 1) in the A/D-conversion procedure, 2) to present the digitally stored measurement on a CRT-display. The D/A-converter has a resolution of 8 bit.

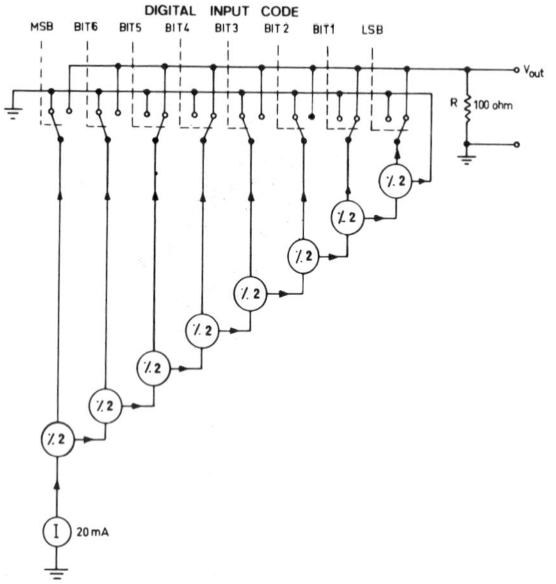

Fig. 3 D/A-converter Principle

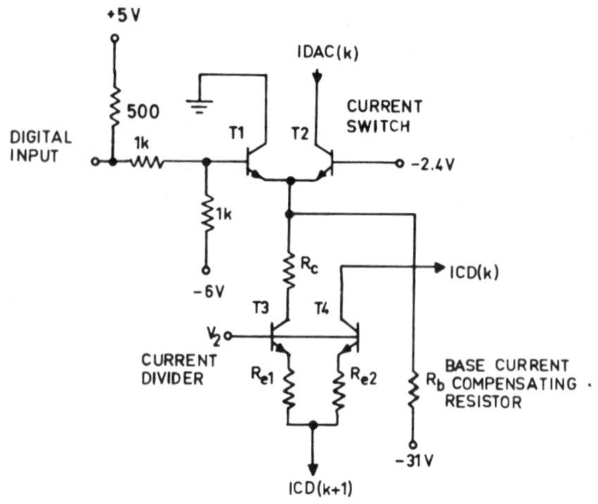

Fig. 4 Current Divider and Current Switch

In the first application the D/A -converter must perform 8 conversions for each sample. Hence the converter must be of very high speed. About 60 ns of the system cycle time can be used by the D/A-converter. To obtain the desired speed the converter is operated in the current mode. The principle is well known: a current is divided into eight binary-weighted currents. The currents, which correspond to the digital input code are then added to give the output voltage. See Fig. 3.

The division of current branches is made by differential pairs of closely matched transistors, Fig. 4. The selection of the current components is made in a second stage of differential transistor pairs. The selected currents are then added in a resistor of 100 ohms. The voltage across this resistor is compared with the analogue voltage that is to be digitized and is also used for the display. The four transistors in the current switch and the divider are on a common monolithic substrate. They are thus held at the same temperature and have closely matched parameters.

The current mode switch, which is used in the D/A-converter, has fast switching times over a wide current range (10 mA - 0.08 mA) with the transistors used. However, the accuracy is low due to the finite current gain of the transistors. Therefore, the switch must be compensated. This is done in a final trimming, where the voltage across the summation resistor for each bit is adjusted to its correct value by a separate resistor R_b. Through this resistor a small amount of extra current is injected into the emitters of the current switching pair.

The D/A-converter has an operating temperature range of -5°C to +85°C. The settling time to within 3.5 mV (1/2 LSB) of the final value is about 50 ns for the prototype.

2.5 The Track-Hold Unit (T/H)

During the A/D-conversion period the analogue signal must be held at a constant value. This is done by sampling the signal in a Track-Hold unit. The Track-Hold unit consists of two Sample-Hold circuits, the inputs of which are connected to a 10 times amplifier. See Fig. 5. The outputs of the two Sample-Hold circuits are connected to the inputs of a two-channel multiplexer. The commands to the Sample-Holds and multiplexer are arranged in such a way that one Sample-Hold is sampling while the stored value of the other one is available at the output of the multiplexer amplifier.

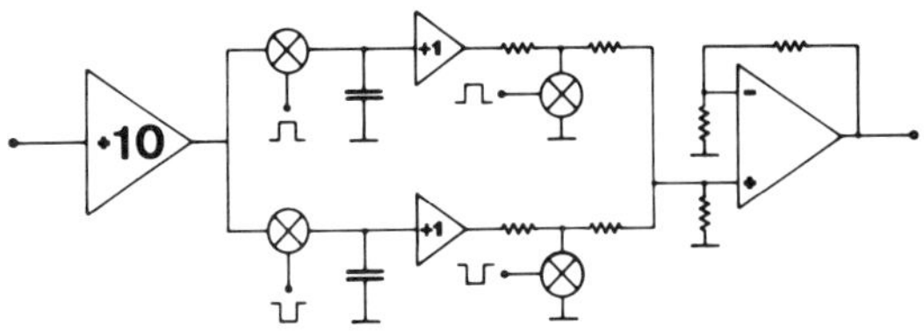

Fig. 5 Track-Hold Unit

The input amplifier has high current capability to be able to drive the Sample-Holds. Each Sample-Hold circuit consists of an analogue switch, a capacitor and a high input impedance voltage follower. Each multiplexer channel consists of a series resistor followed by an analogue switch connected to the ground. The two channels are connected to a summing amplifier via two resistors. The acquisition time for the total Track-Hold unit is about 2 us. The input signal is integrated by a time-constant of about 300 ns.

The multiplexing technique reduces the requirement for high speed Sample-Hold circuits which means that the unit can be built at a low cost. It is possible to extend the multiplexing system by connecting more Sample-Hold circuits in parallell. In this way the speed of the entire Track-Hold unit can be increased without the need for faster Sample-Holds.

2.6 The Comparator

The comparator consists of a differential voltage comparator and a Schottky TTL D-type flip-flop. The total propagation delay of the comparator is about 40 ns. The outputs from the D flip-flop are connected to the arithmetic unit command lines for addition and subtraction. The D flip-flop is strobed not less than 120 ns after the positive edge of the system clock.

3.1 Control Part

The control part of the processor is organized as a number of independent sub-units. The interaction between these is controlled by "Procedure Control Units" (PCU). This implementation may be compared to a software program organized as a sequence of subroutine calls. Several PCU's can, however, be active simultaneously. The starting of one PCU may depend on logical conditions from several PCU's. Some of the advantages in using PCU's are: (1) fast execution of instructions is possible (up to 10 instructions / us) (2) the design and addition of new functions is simple.

At the present time the system comprises 11 PCU's. Three in the main program: for resetting the memory, for signal averaging, for display of the memory contents. The other PCU's are subordinate. The following three paragraphs describe the PCU and one procedure and also the implementation of the signal averaging measurement program.

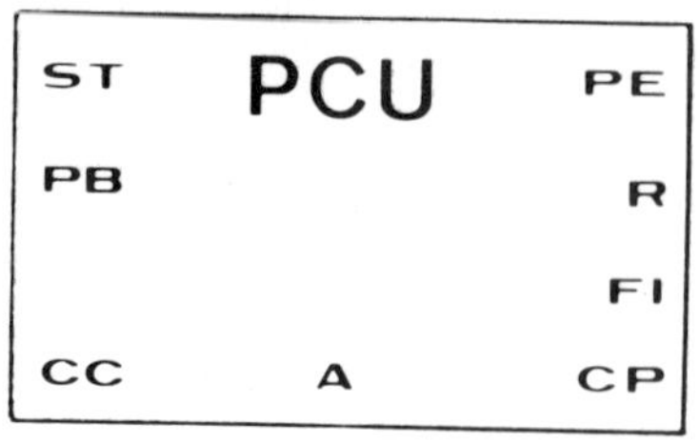

Fig. 6 Procedure Control Unit

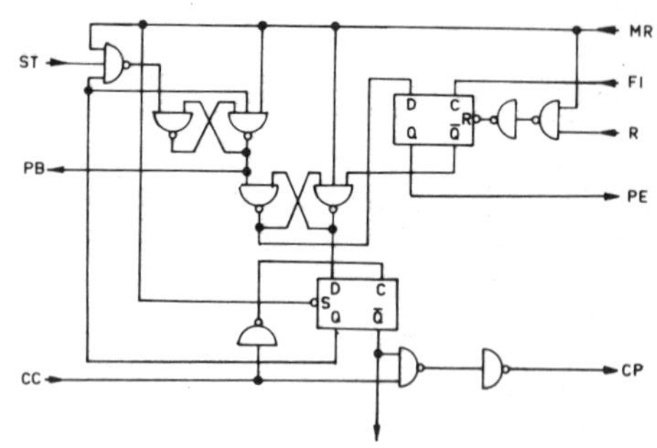

Fig. 7 Logic Implementation

3.2 Procedure Control Unit (PCU)

The PCU controls the interaction between the different subroutines. Each PCU activates the logic circuits controlling the corresponding procedure. Figure 6 shows the diagram for the PCU. It has four control inputs ST, CC, R, FI and four outputs PB, CP, PE and A. The operation sequence of the PCU is as follows: (1) the PCU receives a start condition at the ST terminal; (2) the PCU acknowledges this by a signal called procedure begin at the PB terminal; (3) the desired operations are started by conditions at the terminals A (active) and CP (clock procedure); (4) when the procedure operations are completed the PCU receives a finish signal at its FI input; (5) the PCU then passes control on to the next PCU via its output terminal PE (procedure end); (6) not until a receipt signal is received at the R input the PCU is released and can be started again. The CC input is the central clock input. MR is a master reset input which is used to reset the PCU when the power is turned on. Figure 7 shows the logic implementation of the Procedure Control Unit.

3.3 The A/D-conversion Procedure

This procedure performs an analogue-to-digital conversion in about 2 us using the successive approximation method. The arithmetic unit, the comparator and the D/A-converter participate in the procedure. About 20 instructions are executed to perform the conversion. When the Track-Hold unit has taken a sample the A/D-conversion PCU is started. The A/D-conversion procedure with a PCU and random logic is shown in Figure 8.

The binary word $1000\ 0000_2$ is loaded into the T-register and transferred to the A-register. The contents of the A-register is converted to analogue form by the A/D-converter and compared with the analogue sample. In the meantime the bit in the T-register has been shifted one step to the right. If the comparator finds the sample larger than the A-register value the bit in the T-register is added into the A-register. If on the other hand the sample is smaller than the value of the A-register the corresponding subtraction is performed by means of the 2's complement method. The T-register is shifted one step to the right for every cycle. When the least significant bit of the T-register has been adjusted in this manner the A/D-conversion procedure is finished.

3.4 The Measurement Procedure

The hardware implementation of the measurement program is shown in Fig. 9. The program is started with a manual start button. If the PCU for the display procedure is active, this procedure is first finished and then the measurement PCU is started. This PCU controls the sweep counter and the sweep PCU. The sweep counter is preset manually to the desired

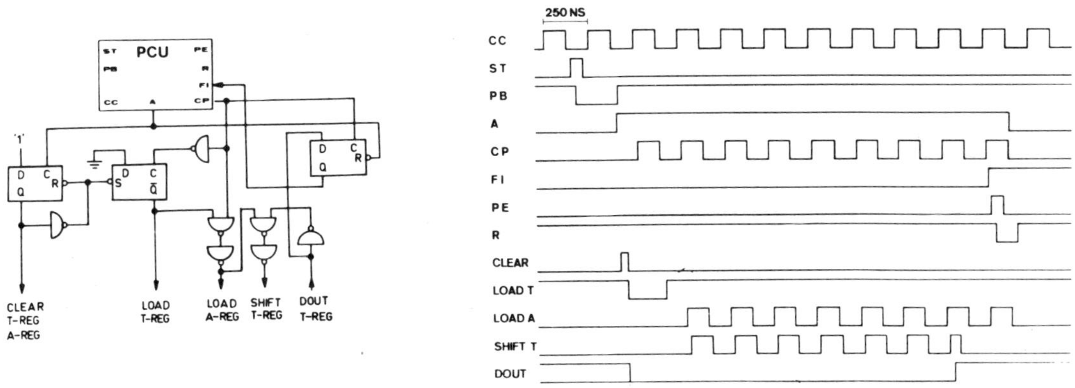

Fig. 8 A/D-conversion Procedure

number of sweeps. The start of a sweep is synchronized by the sweep PCU which activates the point counter and the point PCU. The point counter is a 256 counter and it is connected to the address part of the memory. The point generator, an external generator with a maximum frequency of 200 kHz, starts the point PCU which activates the A/D-conversion PCU and the read PCU. The new digital value for the sample point is added to the corresponding value from the memory during the add procedure. The result is written into the memory during the write procedure. The write PCU stops the point PCU and increments the point counter. The point PCU is then ready to receive a new start command from the point generator. This sequence is repeated for 256 sample points. Then the point counter stops the sweep PCU and increments the sweep counter. When all the sweeps have been done the sweep counter stops the measurement PCU. The display PCU is then started by the measurement PCU.

4. Conclusions

The recent advances in IC technology have made it possible to construct the instrument at a low cost. This has been done using a time-shared D/A-converter, by programming the A/D-conversion procedure instead of using a separate A/D-converter and by the development of a high speed D/A-converter and a Track-Hold unit for this application. 177 TTL DIP logic, mostly gates and flip-flops, are used in the instrument.

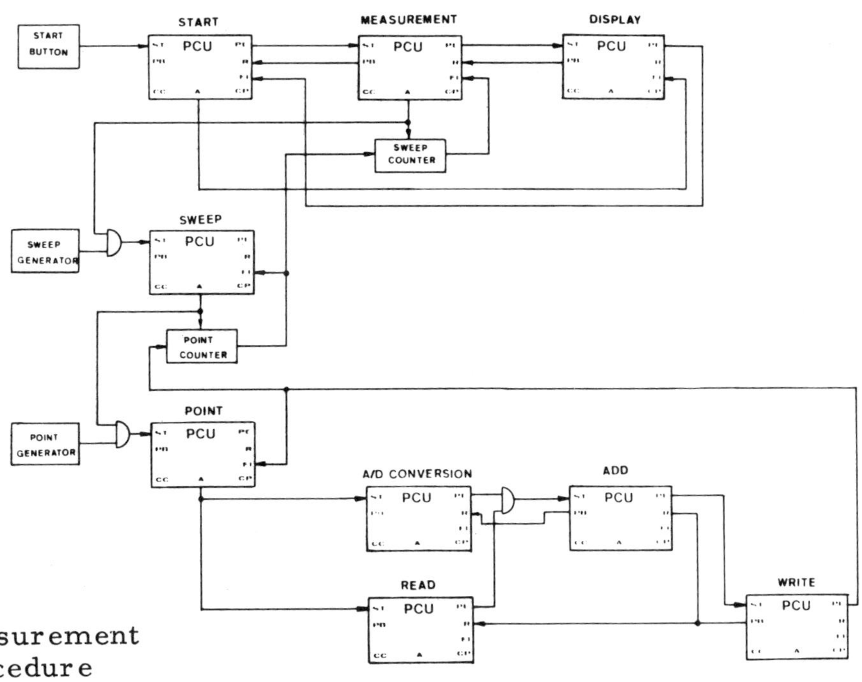

Fig. 9 Measurement Procedure

A DIGITAL ANALYSER FOR REAL TIME STATISTICAL COMPUTATIONS

C.J. Harris and C.T. Thomas

Introduction:

In many control and instrumentation systems, measurement of statistical characteristics such as moments, correlation and spectral analysis is becoming increasingly significant. In particular in mechanical vibration studies, state estimation and adaptive filtering, system identification and information processing.

In this paper a digital instrument is described that operates in real time, to compute the first four time average moments of a randomly varying voltage $v(t)$. The instrument can be easily extended using much of the same logic circuits and storage facilities to evaluate the autocorrelation function and the cumulative amplitude probability density of the input signal v. A statistical signal may be completely described by its moments, from which the mean, median, variance, skewness and kurtosis may be deduced. These parameters, given by the first four moments, almost completely characterise the random variable since they approximately define the probability density of v. Higher order moments are only necessary to describe limiting properties of the random variable.

The input analogue signal v is sampled by an equi-interval seven bit word successive approximation type A/D converter. The sampling frequency varies between 1 Hz and 100 kHz, allowing the analysis of input signals up to 16.5 kHz with 99% confidence of the error being within 1%. At the end of the measurement time, all measures are available either as displays with sign digit or in coded form for direct interface with ancillary equipment for applications such as adaptive filtering or estimation. The complete system cycles over 32 clock beats - the first ten are used to sample and convert it to a seven bit word plus sign bit, the remaining clock beats are used to compute the moments via an arithmetic overflow technique coupled with weighting factors that avoid the problems of excessive processing time and the large number of components of conventional multipliers. All computations of a sample are completed within the 32 clock beats - avoiding difficulties of programming and unnecessary storage facilities. Display of the moments, autocorrelation function (with time delay) and measurement time is by five digit decimal indicator tubes. Evaluation of the autocorrelation function is made by using the same basic circuitry of the second moment with delay and sample storage being controlled by the system clock generator. The principle of operation, theory and system design of the statistics processor are discussed in detail in the following sections.

C.J. Harris is with the Control Systems Centre, UMIST

C.T. Thomas is with the GPO, formerly with Hull University, Dept. of Electronic Eng.

Digital Real Time Moment Algorithms

For a random signal $v(t)$ that is ergodic with density function $p(v)$, its rth moment is given by

$$m(r) = E[v^r] = \int_{-\infty}^{+\infty} v^r . p(v) . dv = \lim_{T\to\infty} \frac{1}{T} \int_0^T v^r(t) . dt \quad .. (1)$$

which may be written in discrete form as

$$m(r) = \sum_{j=-n}^{n} v_j^{\ r} . p(v_j) \quad .. (2)$$

where $p(v_j)$ is the probability that $v(t)$ lies between the jth positive and negative quantisation levels (n in total on either side of zero). If the voltage at the nth level is unity then writing equation (2) in terms of the midinterval voltages $V_j = \frac{(2j+1)}{2n}$ we have

$$m(r) = \sum_{j=-n}^{n} (P_j - P_{j+1}) . (\frac{2j+1}{2n})^r \quad .. (3)$$

where P_r is the probability that the sample lies above the jth level. Now since the expressions for the positive $(m(r)_+)$ and negative $(m(r)_-)$ parts of $m(r)$ are identical apart from a sign difference, equation (3) becomes, on noting that $P_n = 0$,

$$m(r)_+ = \frac{P_o}{(2n)^r} + \frac{1}{(2n)^r} \sum_{j=1}^{n-1} P_j [(2j+1)^r - (2j-1)^r] \quad .. (4)$$

However, for n large (≥ 6) $\frac{P_o}{(2n)^r}$ is negligible for all r.

For the sampling process $P_j \simeq \frac{s_j}{s_t}$ where s_j is the number of samples above the jth level and s_t is the total sample size in the measurement time T, then (4) becomes after simplification,

$$m(r)_+ \simeq \frac{1}{n^r s_t} . \sum_{j=1}^{n-1} (r\, j^{r-1}) . s_j \quad .. (5)$$

for $r = 1,2,3,4$. An identical equation holds for $m(r)_-$.

Implementation of equation (5) requires the computation of $(r\, j^{r-1})$ for each sample exceeding the jth level and the addition of this number to an accumulator. This is wasteful in hardware and the processing time would far exceed the sampling period of the input analogue signal. An alternative is to use a weighting factor[1] for each level and a single comparator, allowing simultaneous consideration of all levels at each sampling instant. It is obvious if the jth level is exceeded then all levels 0 to $(j-1)$ are also exceeded at that sampling time. Then a precomputed weighting factor $WF_r(j)$, can be made for each rth moment at the jth level, such that

$$WF_r(j) = \sum_{s=1}^{j} r\, s^{r-1} \quad .. (6)$$

In particular, $WF_1(j) = j$, $WF_2(j) = j(j+1)$

$$WF_3(j) = \frac{j(j+1)(2j+1)}{2} , \quad WF_4(j) = j^2(j+1)^2 \quad .. (7)$$

Moments Processor Operation

A block diagram of the complete system is shown in fig. 1, and a flow chart of the processor is shown in fig. 2 . The digital system operates on positive voltages, consequently the input signal is rectified. Conventional rectifiers of diode or switching amplifier design were found to be unsatisfactory and a level changing circuit was incorporated in the feedback loop of the A/D converter; this ensures that an input zero potential coincides with the midrange value of the converter. The first or most significant bit (MSB) from the converter represents the polarity of the sampled input signal, and successive digits its amplitude. For negative samples - represented by '0' in the MSB, the complement of the remaining binary word is taken to give a ones complement representation. This is convenient in the calculation of the moments by the arithmetic overflow method. A successive approximation type analogue to digital converter is used, since it is considered superior in accuracy and speed of conversion compared with ramp techniques, and preferable to parallel conversion techniques which are costly in components and of low relative accuracy. For a n bit word, conversion is achieved in n iterations, a 7 bit word conversion was adopted giving 128 voltage quantisation levels for the desired accuracy. The complete system is controlled via a 3.2 MHz crystal clock, each set of computations per sample taking 32 clock pulses, the first ten are used for sample and hold and analogue to digital conversion; the remaining pulses are used for moment generation, intermediate register clear, and sample count increment.

Scaling down to a range of $\pm$ 1 V of the input analogue signal is provided at the input stage by a multiswitch-resistor circuit which is coupled to the moment generators and the output displays to ensure correct moment value accumulation.

Conventional multiplication of binary numbers requires $2n-1$ clock pulses. However, by using both edges of the clock pulses, multiplication can be performed in n pulses. Shift, Add and Store operations are performed with conventional logic. The main disadvantages of this approach is that calculations must be performed in binary form, which is not suitable for decimal display or in BCD.

For the weighted feed logic representation (equation (6)) for the moments $m(r)$, multiplication is achieved simply by determining the highest level exceeded at a sampling instant and then feeding the corresponding precomputed weighting factor into an accumulator. For even order moments the same logic and accumulators are used for $m(r)_+$ and $m(r)_-$. If r is odd, then the weighting factor is computed by the same logic but separate accumulators are used for $m(r)_+$ and $m(r)_-$, and the accumulation process is controlled by the sign digit of the sampled value of the input analogue signal. These accumulators feed UP/DOWN counters controlled by the sign digit. For the rth moment, the accumulator is required to store ik bits where $2^i = n$. Carryout or overflow of the accumulators occurs when the number of pulses accumulated is greater than or equal to 2^{ik} . The overflow represents the contribution to $m(r).s_t$ of that sample. The main advantage of this method is that counting can be performed in any code. The main disadvantage lies in the large amount of logic circuits necessary to evaluate the weighting numbers for each moment.

An alternative approach in the generation of odd order moments is by the arithmetic method of ones complement coding of negative numbers, which allows addition and subtraction to be carried out by binary addition only. Basic to this method is the end round carry digit, caused by overflow, which is added to the least significant digit of the resultant binary word. The disadvantage of this method is the output is only available in binary form. In the instrument described in this paper, both methods of moment generation are combined to give the arithmetic overflow method[2].

In operation the processor has four set variables, these are: the input analogue signal bandwidth, the associated sampling rate, input analogue amplitude range, and the sample count (1 to 10^4 samples with manual override) or integration time. Each of these variables has to be set and/or known for operation of the processor.

Moment Generation Circuits

For the first moment, the processor forms the sum of the samples taken from the input analogue signal; in this capacity it operates simply as a Digital Differential Analyser. Samples are available in ones complement from the A/D converter, each signal is summed with the output from the binary register in addition to having 'end round' carry connections (fig. 3). The MSB is logic monitored and modified should overflow occur. The polarity of the output BCD register is indicated by a polarity bistable. Should the BCD counters output be zero the polarity bistable is set by the incoming signal. A zero output sensing bistable is included to gate the up-down counter. An exclusive-OR gate detects overflow in the full Adder and gates the input to the synchronous BCD counter. A clear display and overflow bistable reset is incorporated. The first moment generator is controlled by three pulses, the first sets the polarity bistable to the sample sign, the second pulse gates pulses to the output register if overflow is present and the third pulse updates the binary register.

Of the 32 control pulses the first ten pulses P_1 to P_{10} are used in the A/D converter, pulses P_{11} to P_{21} are used to evaluate the first and second moments. The weighting factor for the second moment is given by equation (7), the polarity sign being of no significance for even order moments, and is computed by a simple multiplication procedure based on the shift and add principle:- for brevity the description of the second moment generator is omitted. The third moment (fig. 4) is more complex to generate, since it has associated polarity. Its weighting factor is given by

$$WF_3(j) = WF_2(j) \cdot \frac{(2j+1)}{2} ,$$

which involves the arithmetic overflow method of the first moment together with the multiplication algorithm. $\frac{(2j+1)}{2}$ is generated in ones complement form and is adopted as the multiplicand, $WF_2(j)$ is taken as the multiplier. Control pulses P_{20} to P_{31} are used to induce an add and shift operation. In response to the negative clock edge the monostable of fig. 4 produces an output pulse of 30 nsec duration, which gates the count pulse to the BCD registers. The trailing edge of this pulse is also used to add the second moment data into the binary register. After 15 nsec delay, this same pulse causes a shift action. The control pulses P_{20} to P_{31} also gate the polarity bistable with the sample sign.

The fourth moment's weighting factor is simply the square of the second moment and a description of its operation is omitted. Third and fourth moment generation is only possible after the complete generation of the second moment at P_{18}, consequently the fourth moment generation utilises pulses P_{20} to P_{31} .

The Control Circuitry and Outputs

The primary function of the control circuitry (fig. 5) is to direct the sequence of operations in the processor: specifically, to initiate the A/D converter, compute and control moment generation, update output and display registers, and increment the sample count. Owing to the wide-band input analogue signals, a variable sampling frequency coupled with an automatic sample count is desirable to permit the machine to have input-output interfacing with other digital systems for automatic on-line operation. A highly stable 3.2 MHz crystal controlled oscillator is incorporated in the control unit, giving an associated upper sampling frequency of 100 kHz. Variation of the sampling rate is provided by five decade counters. Start and Stop facility of the processor is achieved respectively by setting a '1' into the serial input of the shift register and by breaking the feedback loop via a RS flip-flop.

The sample count circuit is constructed from modulo 10 counters in series, their outputs are manually selected to feed the reset bistables. A 32 bit shift register, driven by the oscillator produces the control pulses or beats P_1 to P_{32} . The control shift register provides only pulses P_{12} to P_{32} , since the A/D converter operates independently of the control block, cycling at the 3.2 MHz clock frequency.

Since the outputs of the moment generators are formed by arithmetic overflow, they can be coded in any form, then BCD moment outputs are used to drive "Nixi" indicator tubes via integrated circuit driver decoders.

Conclusions

Considerable work on the theoretical approach to error analysis of the moment generation process by the weighted feed logic approach has been made[2,3]. The major errors arise from amplitude quantisation - for a seven bit A/D conversion this represents a maximum error of 2% for a normally distributed input signal, and from using a finite sampling period, which gives a worst case error of 3% for a 50 kHz input sinusoid. In normal operation these errors are of the order of 1%.

REFERENCES

1. Kitai, R., Masuko, A.: Digital instrument for measurement of autocorrelation and moments. Proc. IEE, Vol.116, No.11, Nov., 1969, pp.1950-1956

2. Thomas, C.T.: A Statistical Moments Processor. Dept. of Electronic Eng. Tech. rpt. Hull University, June 1973

3. Deist, F., Kitai, R.: Digital transfer voltmeters: principles and error characteristics. Proc.IEE, Vol.110. No.10, Oct. 1963, pp. 1887-1904

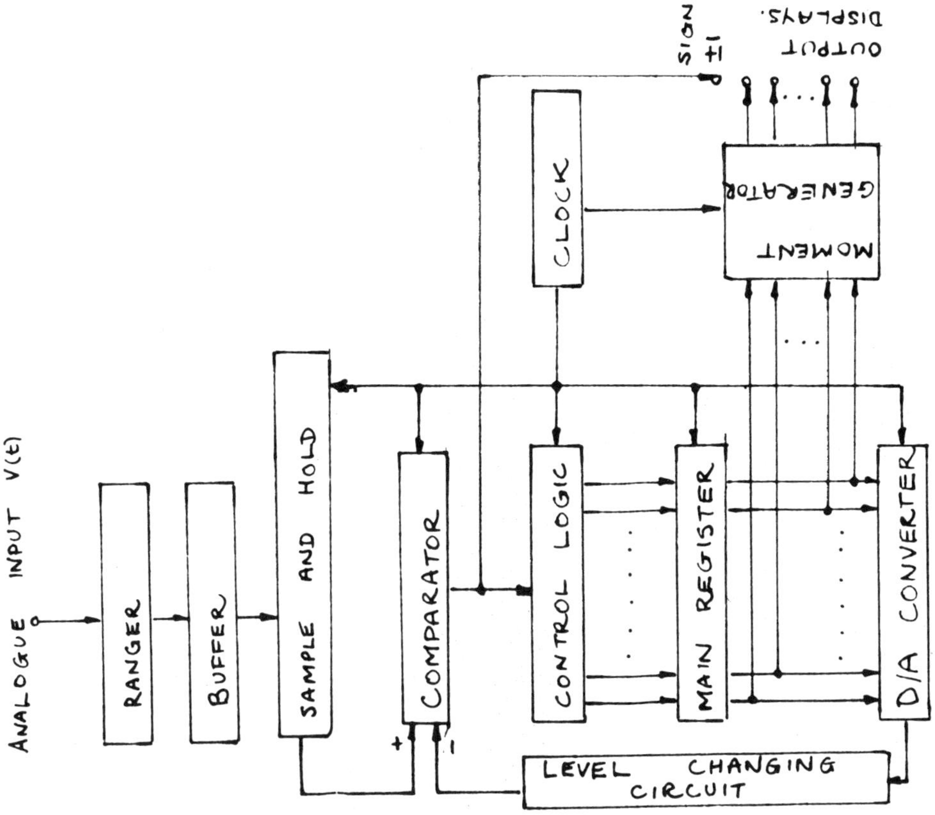

Fig. 1 Block representation of moments computer.

Fig. 2 Flow chart of moments computer.

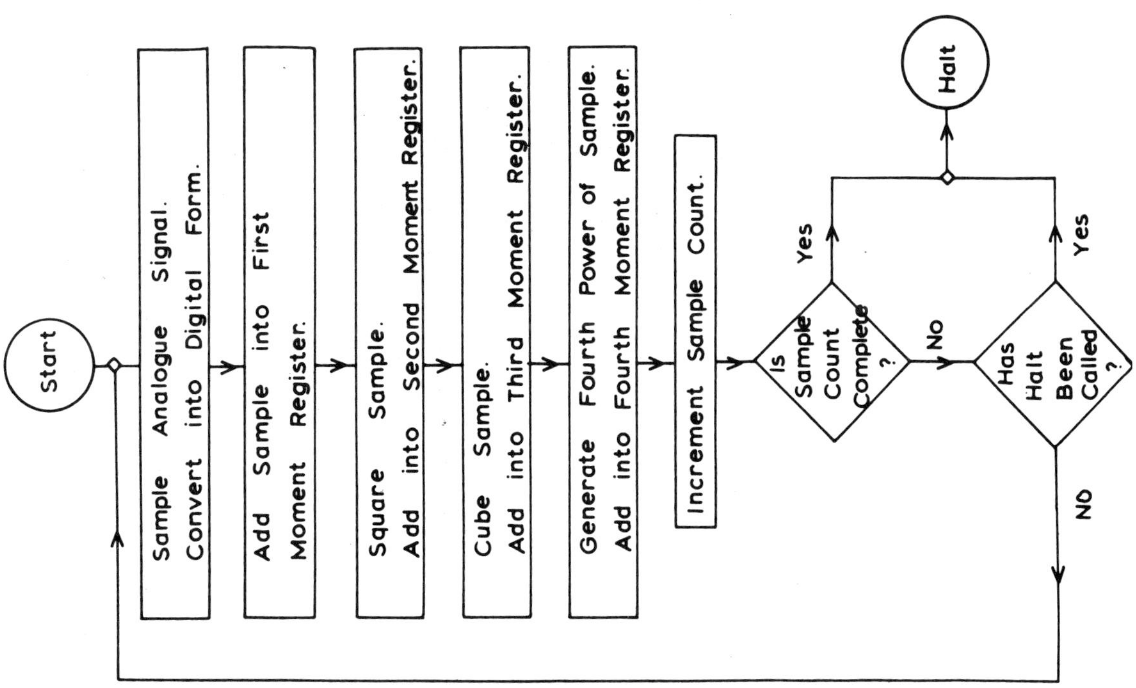

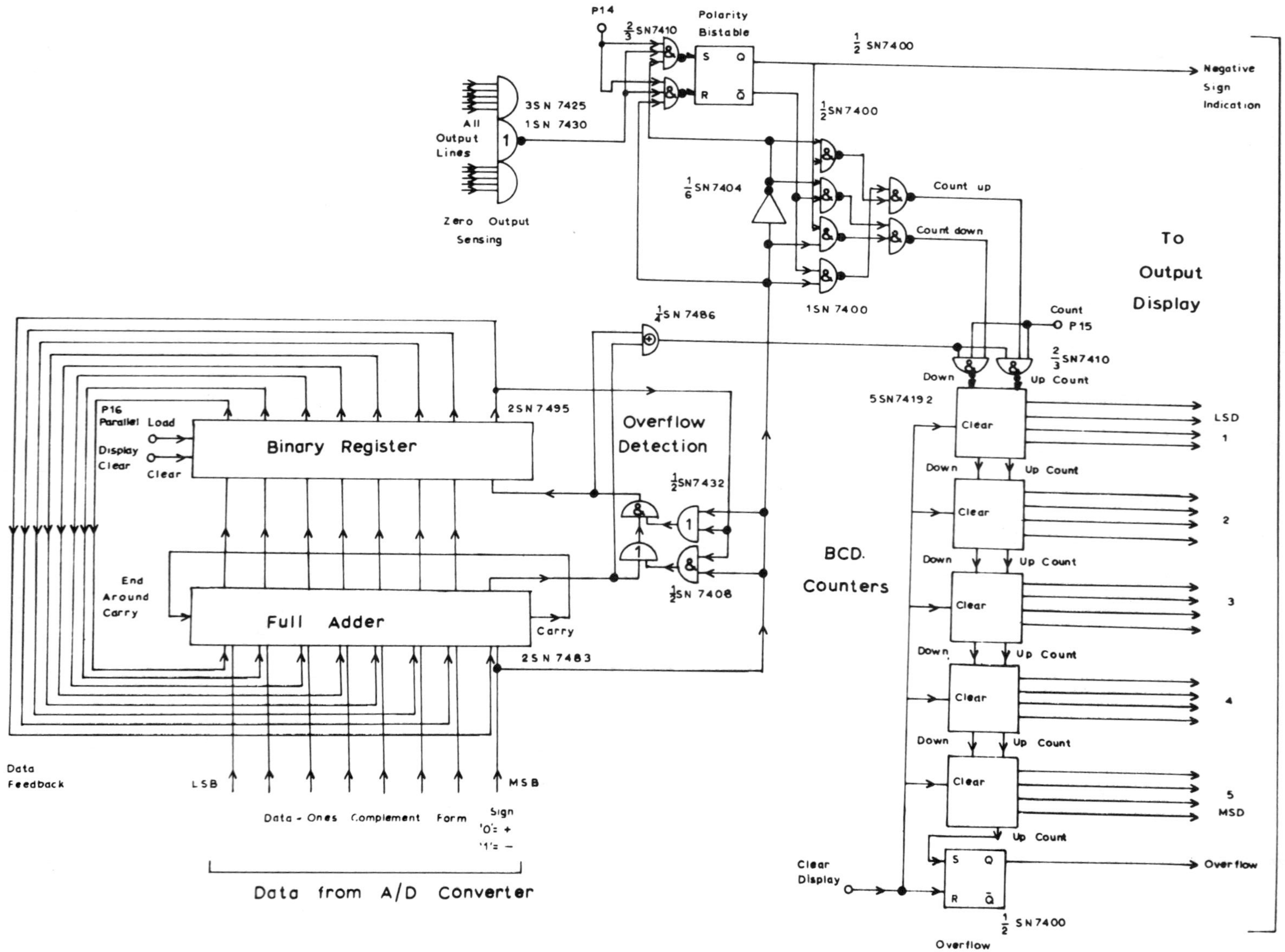

Fig. 3 First moment generator circuit

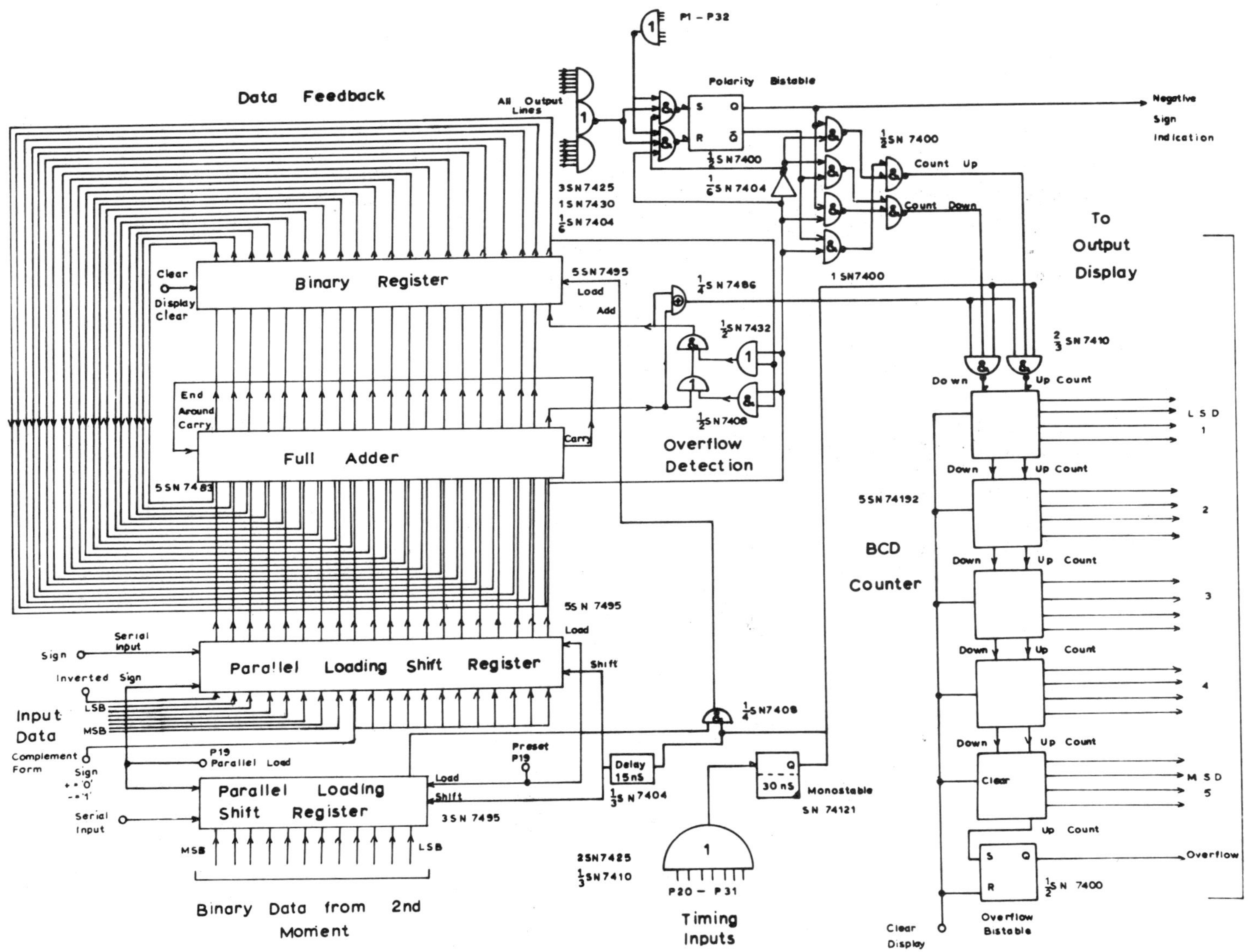

Fig. 4 Third moment generator circuit

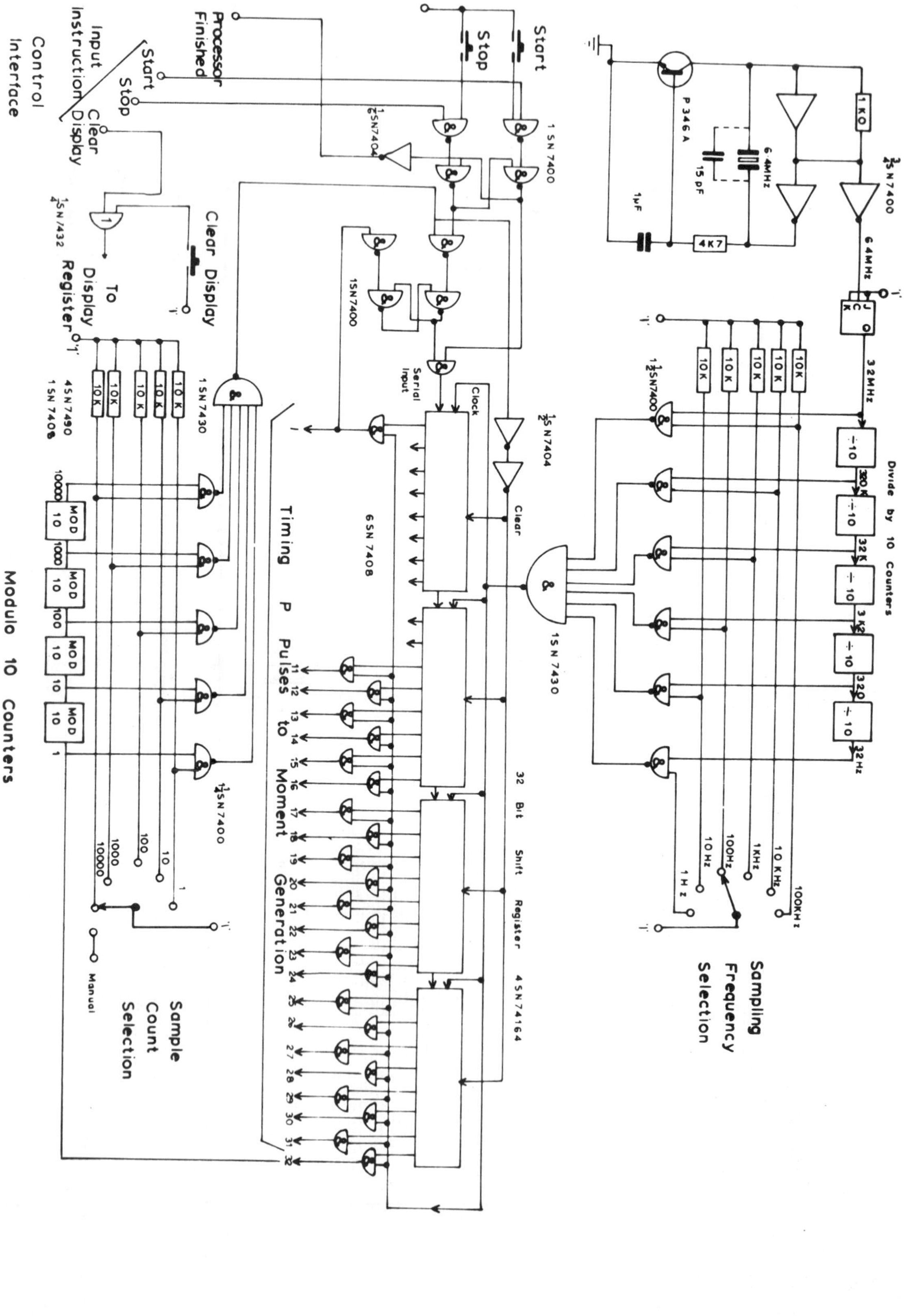

Fig. 5 Processor control and clock circuitry

GENERATING AND CORRELATING PSEUDO-RANDOM BINARY SEQUENCES AT 1 GIGABIT/SECOND

J.A. Coekin and J.R. Wicking

Introduction:

As the demand for data transmission over national networks increases, authorities responsible for the long-term planning of communications are looking ahead to the implementation of digital communication networks in the range 100-800 Mbits/s. Consequently methods must be established for the testing of such systems and since their purpose is to carry digital information what could be more appropriate than to determine their impulse response - a procedure referred to as "identification" since it yields the system transfer function. Direct impulse testing is notoriously difficult and unrewarding because large impulses may cause saturation and overload, and small ones are undiscernable from the system noise. Impulse response is therefore invariably determined by using correlation methods, in which noise is used as a test signal. The basis of the method is that provided the autocorrelation function of the noise is an impulse (i.e. its bandwidth is much larger than the system bandwidth) then the cross-correlation of the system output with the test signal will be the impulse response of the system. In practice the procedure therefore requires a suitable noise signal $x(t)$ which excites the system to give an output $y(t)$; a variable delay to which the noise input is applied, producing $x(t-\tau)$; a multiplier yielding $x(t-\tau)\,y(t)$; and an integrator to average the product over a long period. The cross-correlation for each value of τ, thus obtained, is equal to $\sigma_x^2\,h(\tau)$ where σ_x is the rms value of the noise input and $h(\tau)$ is the response of the system for delay τ. The output correlogram is constructed from the individual points obtained for successive values of the delay. White noise is not a suitable test signal because its complete randomness produces unrepeatable results and requires inconveniently-long integrating times. Pseudo-random binary noise in the form of bit sequences (PRBS) is, however, a very satisfactory substitute since its characteristics are deterministic although it appears random to the system, and the averaging time can be made equal to the sequence length (which is usually very long for correlation studies) thus producing a repetitive correlogram for a stationary system.

An example of an application is a 100 Mbits/s system, and the condition that the noise input must have an impulse autocorrelation dictates at least a 1 Gbit/s PRBS. Until relatively recently the generation of sequences at such high bit rates was only possible using high speed switching devices (such as

J.A. Coekin is Associate Professor in Charge

J.R. Wicking is Senior Tutor

Electrical Engineering Division, James Cook University, Queensland, Australia.

the tunnel diode) in special configurations - often with complex driving waveforms. But an algorithm has now been formulated (Refs.1,2) which facilitates the use of high-speed integrated circuits, as will be described. The second main problem in implementing cross-correlation studies at these bit-rates is that it is not possible to go on indefinitely increasing the bandwidth of some components of the conventional correlator, such as the multiplier. In fact it would be highly desirable to remove the pressure for increased bandwidth from as many components as possible so that PRBS may be used at higher than 1 Gbit/s. A new form of correlator will be described which largely satisfies this requirement.

High-Speed PRBS Generation.

There are many forms of PRBS but the "m" (or "maximal length") type is most commonly used because it is readily generated using shift registers; has maximal length in that every logic combination of the generating register, except all ZEROS, is used; its autocorrelation is impulsive - increasingly so as the number of bits in the sequence increases. Established methods of generation use shift registers with modulo-2 (exclusive-OR) feedback. The logic state of the first stage of the register is determined by the modulo-2 addition of the last stage and one other (the choice is restricted for m-sequences). For example, a four stage register with modulo-2 feedback from stages three and four generates the 15-bit pseudo-random sequence 000100110101111.

The algorithm referred to above shows that if an m-sequence is modulo-2 added to another sequence which is a replica of it but delayed $-(n+1)/2$ bits (where n is the number of bits in the sequence) plus a half-bit (so that the sequences are staggered) then the result is a replica of the generating sequences, further delayed in time BUT AT TWICE THE BIT RATE. This is simply illustrated for the 4-stage register case. Each of the generating sequences must be written with double bit-length to represent the staggering i.e. one is quiescent while the other changes.

Sequence 1	000000110000111100110011111111...
Sequence 2 delayed by - 8 bits	000000110000111...
Modulo-2 addition	100110101111000...

... which is the original sequence but at twice the bit-rate of the generating registers. The staggering is obtained by clocking the registers from a two-stage ring counter driven from a master clock. The initial loading of the second register is determined from the logic sequence of the basic register. The multiplexing concept may be extended to two pairs of registers, as shown in Fig.1, thus achieving a four-fold increase in bit rates; to four pairs for eight fold, and so on.

An important aspect of this algorithm is that in implementing it, the generating registers can be constructed from the highest-speed integrated circuits available and the bit-rate multiplication is then achieved by the modulo-2 adders. The design of special circuits is therefore confined to these adders, the last of which must operate at the full rate of the final sequence. Again the algorithm has an advantageous feature in that because of the staggering, no two transitions in the

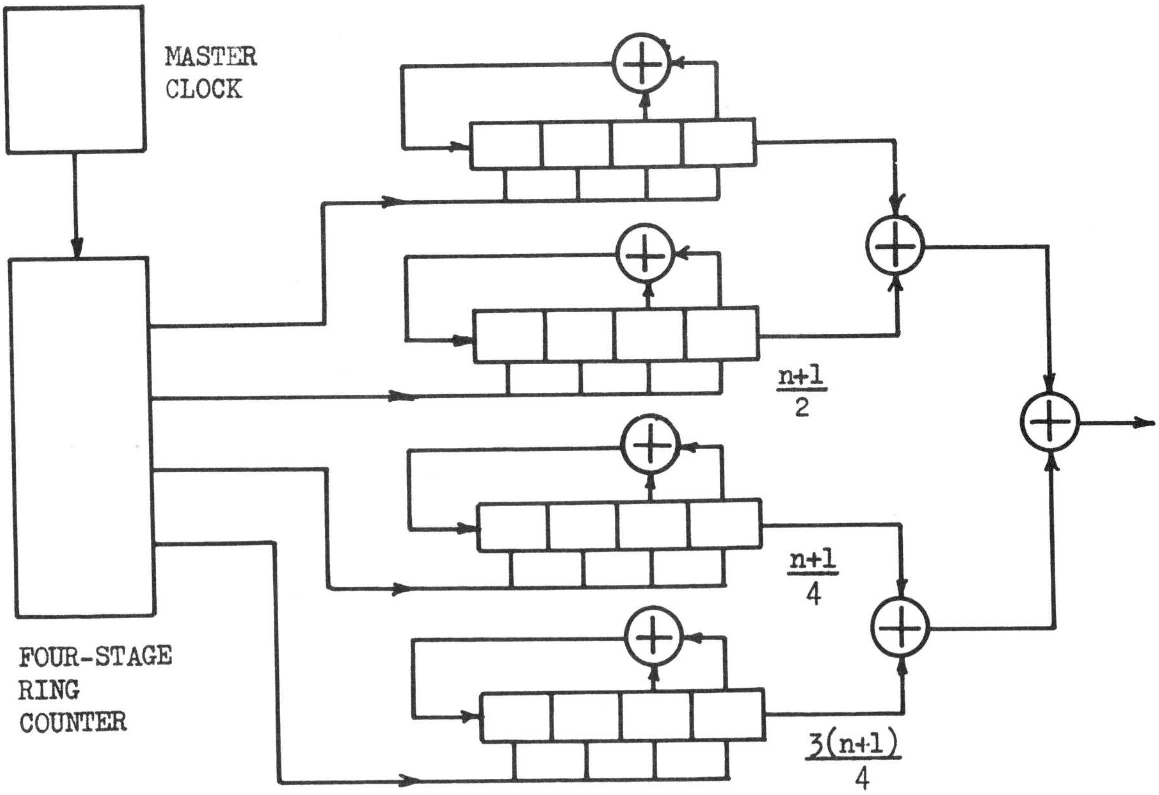

FIG.1. A FOUR-REGISTER PRBS GENERATOR.

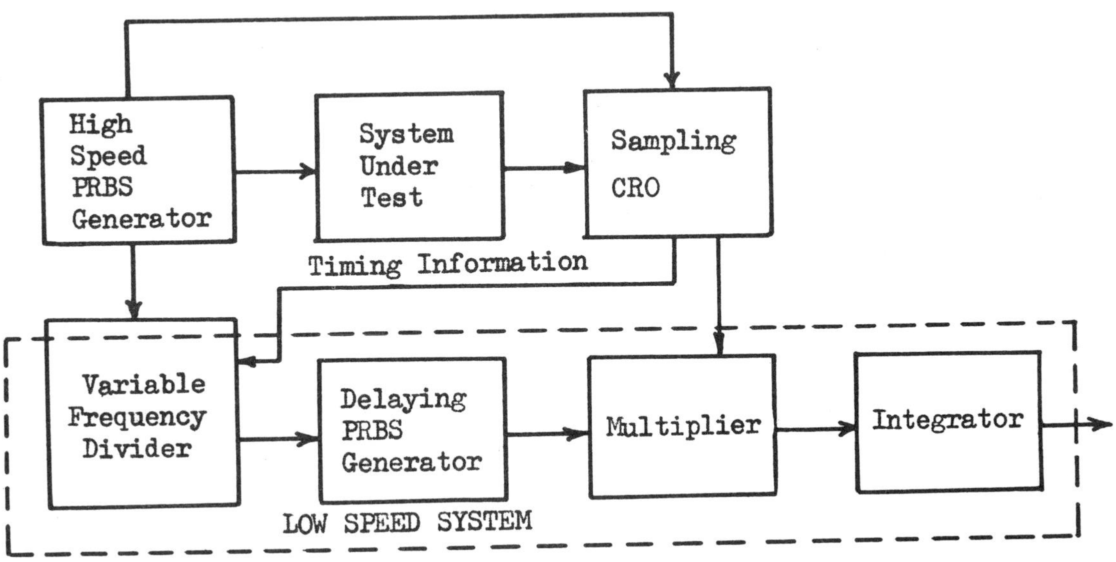

FIG.2. SCHEMATIC OF SAMPLING CORRELATOR

system occur at the same time. The modulo-2 addition can therefore be performed on the transitions rather than on the logic levels, and furthermore provided the transitions are clearly separated in time, only one adder is required to do all the multiplexing. Edge-triggered flip-flops using tunnel diodes have already been developed to operate at 1 Gbit/s and higher and one particular design will trigger on either positive-going or negative-going transitions. To separate the transitions the sequences from the registers must undergo some rise time improvement. Sequences have been produced at about 300 Mbits/s from a pair of registers using MECL III D-type flip-flops and exclusive-OR gates. A toggling MECL III D-type flip-flop was used for the ring counter.

A Sampling Correlator *

The increased bandwidth requirements may be alleviated by using the sequential sampling facilities afforded by a sampling oscilloscope (Fig.2). Suppose that the response of a system to a 1 Gbit/s PRBS is applied to one channel of such an oscilloscope. The display is a slow-speed replica of the response and on the front panel there is normally available a voltage proportional to the vertical deflection. If this voltage is correlated with a synchronised time-scaled version of the original high-speed PRBS used to excite the system, then the resulting correlogram will have the same shape as if the high-speed signals had been used. Cross-correlation can thus be achieved using low-speed circuits, but consequently at the expense of increased processing time.

The most important feature of the operation of this system is the number of samples per horizontal division, taken by the oscilloscope. The time-base sweep-time is a direct function of the number of samples/division because the sampling rate is approximately fixed. A broad correlator bandwidth necessitates a large number of samples but this means a low speed and long processing time. To synchronise the time-scaled PRBS with the display, the ratio of the real time of the sweep to the displayed time must be known. Fortunately most, if not all, sampling oscilloscopes have a sweep output terminal - so the sweep may be timed - and since the display time is given by the control settings, the ratio may be determined. However, an automatic calculator may be constructed in which it is only necessary to insert the display time manually, the sweep being used to control the operation, thus setting the time-scaling. The variable frequency divider of Fig.2 consists of two parts, the first being the ratio calculator. This is composed of a fixed frequency oscillator driving a programmable divider, the characteristics of which are set by inserting the display time. The output of the divider is applied to a gate which opens when the sweep starts and closes when it reaches its maximum. The pulses allowed through the gate are counted in a BCD counter whose outputs are used in the second part of the frequency divider. Pulses from the high-speed ring counter (Fig.1) are first prescaled and then applied to another programmable divider whose state is determined from the outputs of the BCD counter. The output of the second programmable divider is used to drive the delaying PRBS generator (Fig.2), the bit-

* Patent Applied for.

rate of which is therefore time-scaled in accordance with the settings of the oscilloscope. The best form of delaying generator to use is one with automatic delay-calculation facilities since pulse-slipping types would be difficult to check and synchronise. The circuits required for all these operations are mostly conventional and, more significantly, operate at slow speeds permitting the use of relatively cheap integrated circuits. The synchronisation of the oscilloscope sweep with the high-speed PRBS may be accomplished by detecting the ALL-ONES state of one of the registers.

As stated earlier, correlator bandwidth, sweep time and the number of samples are inter-related, and moreover the sweep time is influenced by the length of the sequence and the bit rate. The calculation of processing time must therefore take all these factors into consideration. A typical example will be examined.

Suppose a correlation is to be formed using a 1000 bit sequence generated at 500 Mbits/s (by a four-register generator), and the correlator bandwidth is to be 500 Mbits/s. The sampling frequency must be 5 Gbits/s. Each of the generating registers operates at 125 Mbits/s so the window between the ALL-ONES detections must be $1000/125 \times 10^6$ equals 8 μs. The nearest greater calibrated setting on the oscilloscope is 10 μs. The sampling interval must be $(5 \times 10^9)^{-1}$ so the number of samples/window is 5×10^4. The sampling rate of the oscilloscope is 15 kHz so the time for all the samples is 3.33s. Thus the total processing time is 1000 x 3.33s which is equal to 55 mins. Since the window is 10 μs wide it contains 5 sequences. Suppose the window were reduced to one sequence length. The number of samples would now be 10^4 and the total processing time 11 mins. which is the minimum possible time when the sequence bit-rate and the bandwidth are the same. When a 1 Gbit/s sequence is used for a 500 Mbits/s bandwidth the minimum time is 5.5 mins. The reduction of the window must increase the effect of random noise but the rms error only varies inversely with the square root of the number of sequences used (Ref.3). A further reduction in processing time is possible by providing parallel-correlation e.g. a 1023 bit sequence is produced by a 10-stage register and if each stage were provided with its own multiplier and integrator, 10 correlations could be performed during one sweep. Clearly from all the foregoing, the sequence length and correlator bandwidth need to be tailored to the required measurement and the selection becomes a compromise between speed and accuracy.

A Computer-Based Sampling Correlator.

There has been considerable research into the computational correction of data derived from identification procedures so there would appear to be advantage in basing the sampling correlator on a computer.

Two channels of the oscilloscope may be time sampled, A/D converted and stored in the computer memory and since the bandwidth of the oscilloscope waveforms is low, only medium conversion rates are needed. The contents of the memory may subsequently be punched out on to paper tape for system evaluation by a larger computer. It would be difficult to derive a clocking signal to initiate the A/D converter and some method has to be found for synchronising the computer and the oscilloscope. A closed-loop system may be obtained if the computer, via D/A conversion, controlled the horizontal sweep, taking samples of both channels at incremental positions.

Small computers currently available have a basic core memory of 4K which would be sufficient for the present task; the complete installation needing, apart from the teletype, paper tape reading and punching, and both A/D and D/A converters. Programming might occupy 1000 locations so 3000 would be available for samples. Since the results of only one sweep (or less) would need to be stored and A/D and D/A converters operate in times of 50μs and 100μs respectively, only 0.6s would be required for 3000 samples. Since the evaluation would be undertaken by a larger computer there would necessarily be a delay before results were available.

The only alternative is to provide large additional storage for the mini-computer which may not be prohibitive in laboratories already equipped with the basic mini-computer installation.

Conclusions.

Since advanced high-speed optical, waveguide and microwave systems are being investigated it is imperative to evolve suitable test procedures. The means have been described for the implementation of cross-correlation studies at very high bit-rates. As far as the oscilloscope is concerned the theoretical limit on correlator bandwidth is about 18 GHz but PRBS generation will clearly limit the possibilities, depending upon the successful use of ultra-high-speed devices.

Acknowledgements.

The authors wish to acknowledge the support of the Research Laboratories of the Australian Post Office under Contract No. C048861.

References.

1. Lempel A. and Eastman W.L. High-speed generation of maximal length sequences, Trans. IEEE C-20, pp.227-228, 1971.

2. Coekin J.A. and Wicking J.R. Generating PRB sequences for testing at 500 Megabits/Sec. and higher using IC flip-flops. Proc. Int. Microwave Symp. Chicago, May 1972.

3. Hughes M. and Norton A. The measurement of control system characteristics by means of a cross-correlator, Proc. IEE Pt. B, Vol.109, No.43B, p.77, January 1962.

HIGH FREQUENCY VOLTAGE/CURRENT SWITCHING TECHNIQUES RELATED TO THE DESIGN OF MULTI-PATH FILTERS

W.A.Evans* and C.J.Paull**

1.0. INTRODUCTION : Linear networks containing periodically operated switches are variously referred to as digital RC filters, N path filters or commutated networks. Attention has been focused on this class of networks in recent years, largely due to the need to obtain bandpass responses without the use of inductors. Extremely high values of Q factor can be achieved without sacrificing stability. In such a digital filter the values of R and C determine the bandwidth whereas the signal switching or commutation frequency ω_o solely fixes the centre frequency. The resonance curve for a filter composed of single low pass RC sections closely approximates that of a single tuned LC filter; however, a sharper cut-off is realised by cascading two or more sections in each path. All digital RC filters exhibit some form of comb structure response characteristic, and for most applications it is highly desirable to minimise the higher order responses by employing the most appropriate modulating waveforms.

The practical application of these filters is largely confined to audio frequencies. It is here that advantages such as component size and avoidance of switching problems are most apparent. Solid state switches behave in an almost ideal manner with little or no transient breakthrough. Owens (1) has provided a summary of a variety of practical low frequency circuits together with a physical insight into their operation. Recently, Paull and Evans (2) have described the use of a commutated filter to execute the correlation process in a frequency response analyser yielding information about the transfer function of the system under test. It was with this particular application in view that the filters to be described in sections 4.0 and 5.0 were initially conceived.

The purpose of this paper is to examine some of the circuit problems associated with these filters and introduce circuit designs that have extended the upper frequency response of these filters in the experience of the authors. The choice of the modulation function and its effect on the output response is briefly described. Two versions of the filter structure are included - one is based on voltage switching techniques and is capable of satisfactory operation to at least 100 KHz. An analogue switch driver has been specially developed for this purpose. The second circuit is based on current steering techniques, the potential of which has not been fully explored, though the prototype has operated at over 1 MHz.

* W.A.Evans is with the Department of Electrical and Electronic Engineering, University College of Swansea.

** C.J.Paull is with the Department of Electrical and Electronic Engineering, University of Nottingham.

2.0. DESCRIPTION OF DIGITAL RC FILTER : A linear equivalent circuit representation of the filter is shown in Fig.1a. In this schematic the forward signal path is split between four identical parallel channels, each containing a linear multiplier followed by a single pole low pass filter. The outputs of the individual channels are recombined by means of a suitable commutating switch that may, or may not, operate at ω_o.

The operation of the filter may be readily understood by first assuming that the frequency of the signal and phased modulating waveforms of Fig.1b. are identical. The product waveform generated at the output of each multiplier then repeats on successive cycles so that each capacitor acquires a constant charge related to the input signal amplitude and the phase relationship between the input signal and modulating waveforms. When the signal frequency ω differs from ω_o, the charge on each capacitor will vary at the difference frequency $(\omega - \omega_o)$. The capacitor charging rate for each path is controlled by the respective time constant, so that as the difference frequency increases the capacitor voltages must become progressively smaller. In effect, a low pass characteristic is translated into a bandpass characteristic centred about ω_o. The function of the output modulators, in this case simplified diagramatically to mechanical commutators, is to sample the capacitor voltages in sequence, thus generating a stepped output waveform at the second modulation frequency ω_{if}.

3.0. THE MODULATION FUNCTIONS : In all cases to be considered the modulation functions A(t), B(t), C(t) and D(t) are related with $B(t) = A(t-\pi/2\omega_o)$, $C(t) = A(t-\pi/\omega_o)$ and $D(t) = A(t-3\pi/2\omega_o)$.

The choice of modulation function can vary, as shown in Fig.1b.

A Fourier analysis of $A_1(t)$ yields the following expression :

$$A_1(t) = \frac{1}{4} + \sqrt{2}/\pi \{\sin(\omega_o t+\pi/4) + 1/\sqrt{2}\, \sin 2\omega_o t + 1/3 \sin(3\omega_o t-\pi/4) + 1/5 \sin(5\omega_o t+\pi/4) + \ldots \text{etc.}\}$$

The expression contains all harmonics except those with an order that is a product of four. However, the remaining modulation functions are so related in phase as to cancel out d.c. and even order products in the final output. For this case the multiplier may be replaced by a single pole one-way switch operating on the input signal, as shown in Fig.2a. with the dotted section excluded. This is the familiar circuit discussed by a variety of authors. usually with $\omega_o = \omega_{if}$ and referred to as a series N path filter.

A Fourier analysis for $A_2(t)$ shows that :

$$A_2(t) = 4/\pi \{\sin\omega_o t + 1/3 \sin 3\omega_o t + 1/5 \sin 5\omega_o t + 1/7 \sin 7\omega_o t + \ldots \text{etc.}\}$$

The linear multipliers may be replaced with the circuit of Fig.2a with the dotted section included. Single pole two-way switches operate on the direct and inverted signals respectively. The particular combination of elements in each channel constitutes what is generally termed a switched integrator.

$A_3(t)$ is derived from $A_2(t)$ by the inclusion of notches in the basic square wave in such a manner that even, 3rd and 5th harmonics are excluded.

$$A_3(t) = \frac{3.356}{\pi}\{\sin\omega_o t + \frac{2.075}{7}\sin 7\omega_o t + \frac{4.386}{9}\sin 9\omega_o t + \frac{3.97}{11}\sin 11\omega_o t + ..etc\}$$

Finally the 3-level modulation function $A_4(t)$ is shown where :

$$A_4(t) = {}^{2\sqrt{3}}/_{\pi}\{\sin\omega_o t - {}^1/_5 \sin 5\omega_o t - {}^1/_7 \sin 7\omega_o t + {}^1/_{11} \sin 11\omega_o t + ...etc.\}$$

Any harmonic with an order that is a multiple of two or three is suppressed and this constitutes a useful result in practice. Fig.2b. shows an example of how the filter may be realised. If any one path is considered, then the ±1 levels are obtained by setting switches to the same inputs. The 0 level is obtained by switched signal cancellation from the opposing inputs. Further harmonic suppression may be obtained by increasing the number of levels and the consequent hardware complexity.

4.0. VOLTAGE SWITCHED FILTER : (a) Series Switch : The replacement of mechanical switches for frequencies above a few hundred hertz by their solid state counterparts becomes essential, and it can be generally assumed that the switches will be identifiable as a major limitation in the overall performance of these filters. Over a broad frequency spectrum the junction field effect transistor (JFET) is now established as a high performance analogue voltage switch. The transient behaviour of the JFET is illustrated by the switching equivalent circuits of Fig.3(b)(c) where the area under the switch OFF spike is considerably greater than the switch ON spike. In the case of a series RC load, the result is a quasi d.c. error voltage across the capacitor proportional to the switching frequency and given by :

$$V_{offset} \approx \{-V_g R_L C_{gd(off)} + V_g (R_L || R_{ds(on)}) C_{gd(on)}\} f$$

The equation is approximate and does not contain C_{ds}. It serves to demonstrate the existence of a trade off between filter bandwidth ($1/4\pi RC$) and switching offset. Reducing R_L increases the bandwidth but reduces the offset error and enhances high frequency performance. It is not particularly fruitful to analyse more precise models due to the proliferation of inadequately characterised variables which affect performance. The wide variation of junction capacitance with reverse bias is one complication to accurate modelling.

To summarise, the series switched filter suffers from the "floating input" condition to any one path for part of each cycle. Switching to an open circuit yields a large offset error when attempts are made to extend the frequency of operation and is consequently unsuited to this purpose.

(b) The Series-Shunt Switch : In series-shunt operation the load is commutated between signal and ground. Alternatively, the load may be commutated between two or more inputs, as shown in Fig.2(a,b).The significant point is that the load is never allowed to "float". The one pole two-way switch is shown in Fig.3(d).The complementary switching waveforms applied to the gates provide for turn OFF of one JFET at exactly the same time as the turn ON of the other. If it is assumed that C_{gd} remains constant and equal for both devices, then the + ive and -ive switching spikes must cancel at the load. Thus, no high frequency offset is generated,regardless of the nature of the load.

The apparently ideal switching characteristics are degraded, in practice, by circuit and component unbalance. Variations in C_{gd} and imperfect complementary switching waveforms produce the most significant output offsets. In the former case the effective mean offset current is given by $\pm(C_{gd1} - C_{gd2})\ V_g f$. The offset voltage is the product of this current and the circuit resistance to a.c. ground into which it flows. Referring again to Fig.3d, the shaded area on the switching diagram shows the time for which both JFETs are OFF, i.e. $|V_{gs}| > |V_p|$ where V_p is the pinch off voltage. A considerable spike may be generated in this region since the impedance to ground is now solely determined by the load impedance $(R_L >> R_{ds(on)})$. For large signal switching ($|V_g| >$ 10V) this OFF transition period may last for up to 50 nanoseconds even for a fast switch driver. This is by no means insignificant when switching frequencies of 100 kHz or more are contemplated.

(c) The JFET Interface and Switchdriver : The JFET may be switched OFF by any V_{gs} less than V_p, but to obtain minimum ON resistance V_{gs} must be zero or slightly positive, regardless of source voltage. This therefore implies that in the ON condition the gate voltage must be free to follow the input voltage whilst at the same time being immediately responsive to a change in the state of the switch driver. The simple capacitor-diode combination in the gate connection has become well known for its automatic self-biassing action and zero steady state gate current. However, it cannot be used when medium or large a.c. voltages are to be switched since C_g ($> C_{gs}$) tends to ground the gate at high signal frequencies. The result is modulation of the source-drain channel resistance with signal level.

Where the signal level significantly exceeds V_p, the JFET may cut off completely on the positive peaks and an effect somewhat akin to half wave rectification could be observed at the drain. To overcome this, it is frequently recommended that a high value gate source bleed resistor R_1 is inserted, as in Fig.4. Since C_g is quite small (50→ 200pf), this is generally adequate for signal frequencies up to a few hundred kilohertz.

The specially balanced switch driver shown in Fig.4. was designed to provide complementary waveforms for driving the series-shunt switches and reducing offset errors at the higher frequencies. The output waveforms are shown in Fig.3e, where it can be seen that the turn OFF has been slightly delayed with respect to turn ON. This is easily arranged by the simple expedient of introducing small capacitors across R_4 and R_5. The result is the elimination of the transition periods when both JFETs are OFF together. In practice, both JFETs may be momentarily ON together. This form of drive, therefore, simulates the make-before-break action often specified for mechanical rotary switches. In this particular context the nearly constant source impedance seen by the load during switching ensures a minimum of spike breakthrough and any consequent offset error.

Channel filtering is provided by passive two-pole resistor-capacitor ladder networks which, by comparison with the comparable single pole networks, give increased stop band attenuation. The 0.01 μf capacitors are included to suppress high frequency transients. The output commutator constructed using metal oxide field effect transistors (MOSFETs) requires very simple drive circuitry for the 125Hz commutation frequency shown in the schematic. The devices with gates designated R and jX are normally held OFF; however, should it be wished to extract either the in-phase or quadrature component of the signal, the appropriate device may be turned ON, thereby shorting out the outputs from two of the channels and eliminating the unwanted component. This circuit, with careful attention to layout, has performed very satisfactorily at frequencies exceeding 100 kHz.

5.0. THE CURRENT SWITCHED FILTER : The circuit shown in Fig.5 was designed to test the effectiveness of current switching techniques in extending the high frequency performance of a digital RC filter. Analytically, the filter closely resembles the voltage switched four-path filter already described where phased square wave modulation $A_2(t)$, $B_2(t)$ etc. is used at the input to each path. However, the circuit now operates as follows : a differential signal input is first level shifted and then transformed into proportional currents at the collectors of transistors $Tr_{2,3}$. These currents are now routed to the four 1KΩ "current summing" resistors via the transistors Tr_{7-10} connected as current switches and controlled from the T.T.L. compatible modulation inputs. No-signal current cancellation is effected at the nodes e,f,g,h, by the duplicate set of current switching transistors $Tr_{5,6,11,12}$, operating on those currents originating from the negative side of each pair of differentially connected transistors $Tr_{1,2}$, and $Tr_{3,4}$. Thus, the four transistors Tr_{13-16} acting as current sinks, reflect back a negative current into nodes e,f,g,h, and eliminate the voltage pedestal that would otherwise be present. Variations in mark-space ratio in the modulating waveforms do not, in this case, give rise to a spurious no-signal output.

The remainder of the circuit follows the same pattern as the voltage switched filter with the exception of the integrated FET switches used in the output commutator to conserve space. The use of matched components plays an important role in the maintenance of low offset errors in the circuit. The operation of the filter has proved very promising at commutation frequencies in excess of 1MHz, as would be expected, as this is a modest frequency by current switching standards. The absence of glitches in the system has been a particularly beneficial result.

6.0. CONCLUSION : This paper has attempted to relate the interaction between the conception of a specific filter structure and the inevitable circuit constraints that follow its realisation. It is anticipated that the designs portrayed here may find wider application and possibly be included in differing filter structures to adapt to a particular requirement.

7.0. REFERENCES :

(1) Owens, A.R. Digital Filters. I.E.E. Vacation School in Electronic Instrumentation, Lecture 14, April 1973.

(2) Paull, C.J. and Evans, W.A. The Application of a Commutated Filter to the Design of a Frequency Response Analyzer. J.of I.E.R.E.1973, 43.6. 369-378

8.0. ACKNOWLEDGEMENT : The financial and technical assistance in support of this work by the Wayne Kerr Company Limited is gratefully acknowledged.

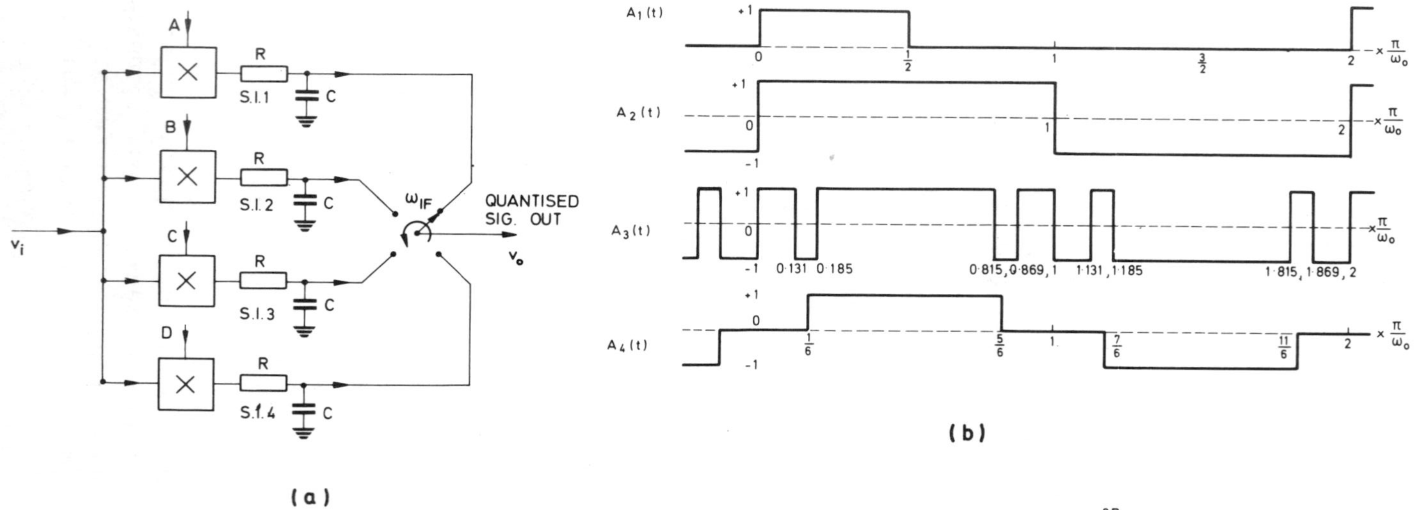

FIG. 1 (a) LINEAR EQUIVALENT CIRCUIT OF DIGITAL R.C. FILTER
(b) MODULATION WAVEFORMS

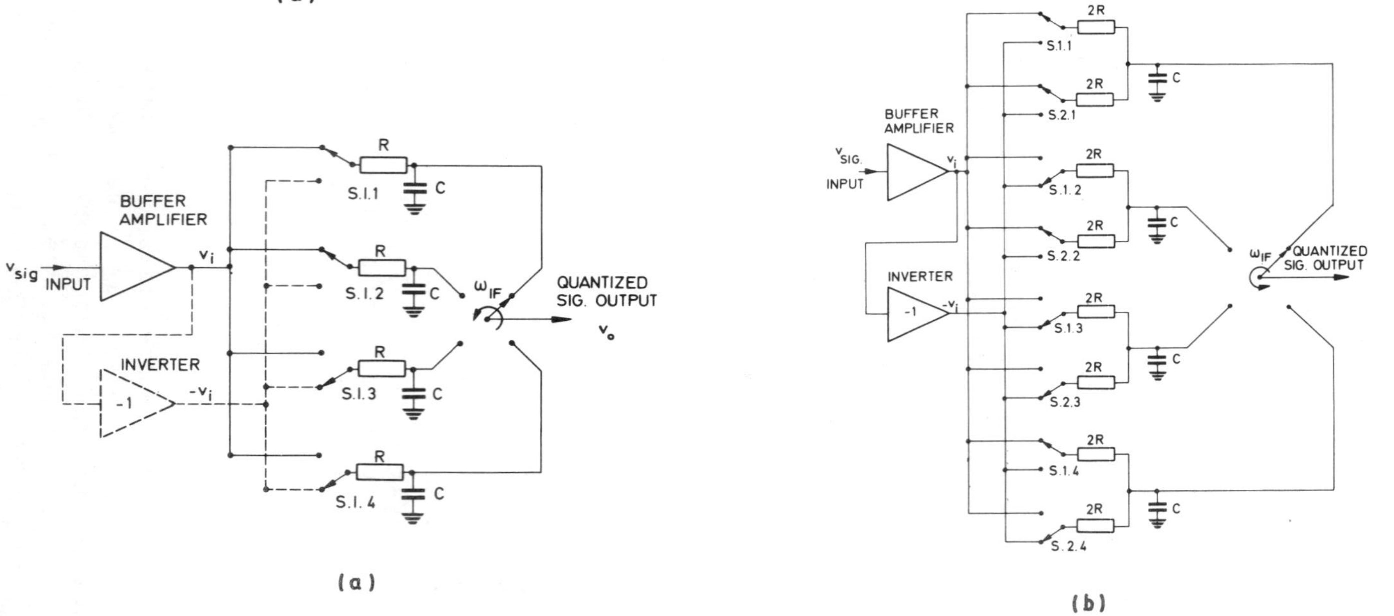

FIG. 2 (a) N-PATH FILTER/ SWITCHING INTEGRATOR
(b) 3 LEVEL SWITCHING INTEGRATOR

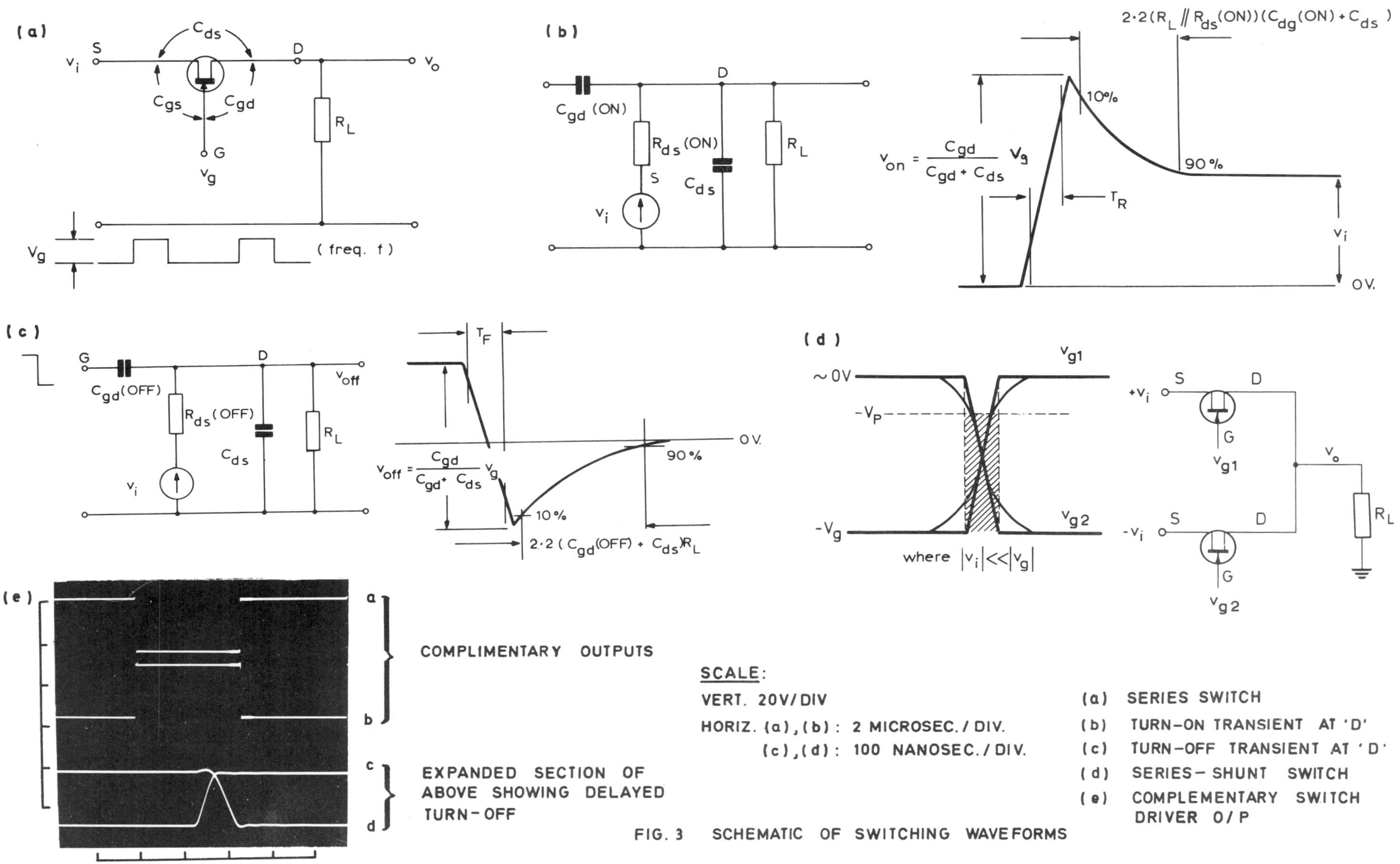

SCALE:

VERT. 20V/DIV

HORIZ. (a),(b) : 2 MICROSEC./DIV.

(c),(d) : 100 NANOSEC./DIV.

(a) SERIES SWITCH
(b) TURN-ON TRANSIENT AT 'D'
(c) TURN-OFF TRANSIENT AT 'D'
(d) SERIES-SHUNT SWITCH
(e) COMPLEMENTARY SWITCH DRIVER O/P

FIG. 3 SCHEMATIC OF SWITCHING WAVEFORMS

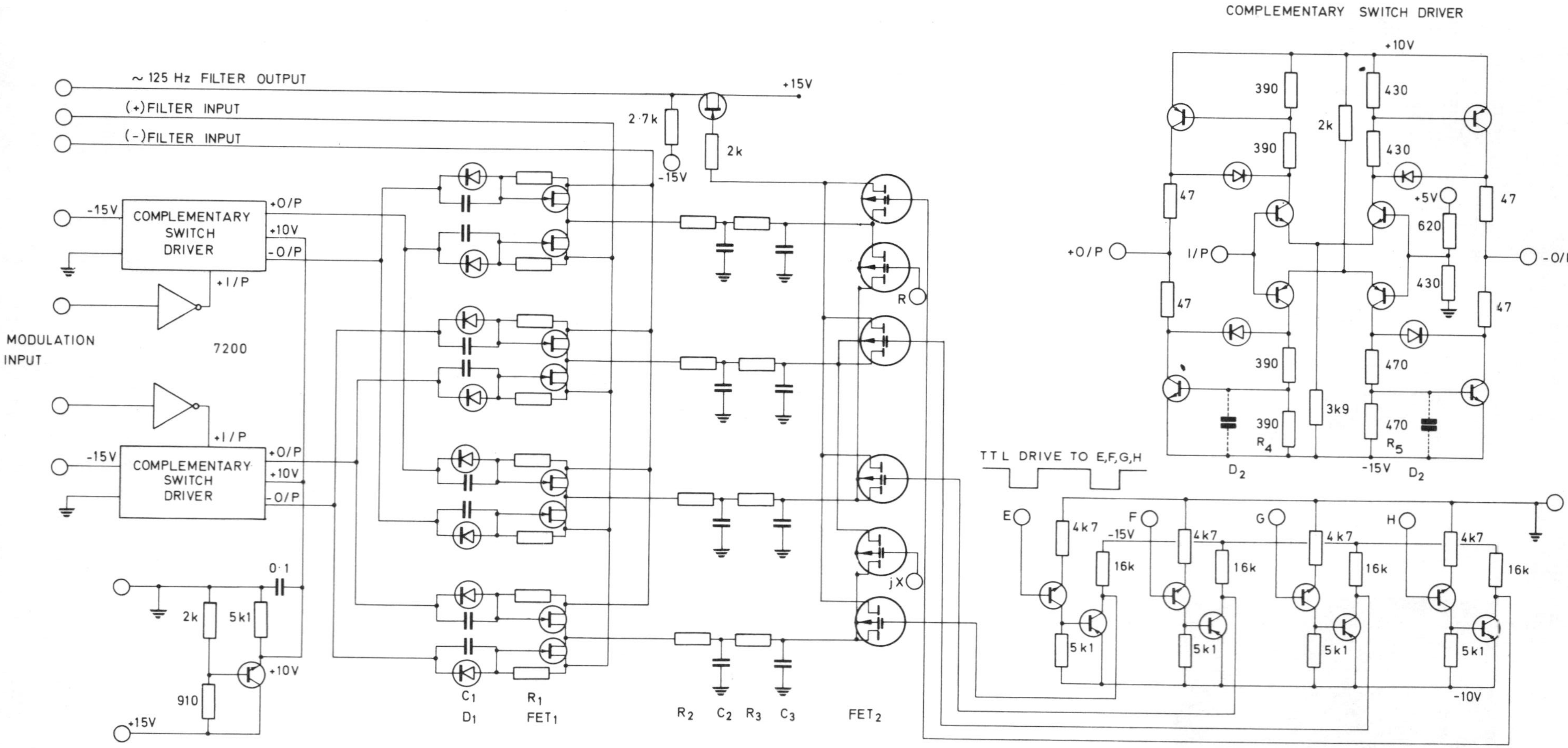

NOTES

C_1 = 47pf , C_2 = C_3=0·47uF
R_1 = 50k , R_2 = R_3 = 220k
NPN = 2N3903, PNP = 2N3905
FET_1 = 2N5432, FET_2 = M1202
D_1 = D_2 = 1N914

FIG. (4) VOLTAGE SWITCHED DIGITAL RC FILTER

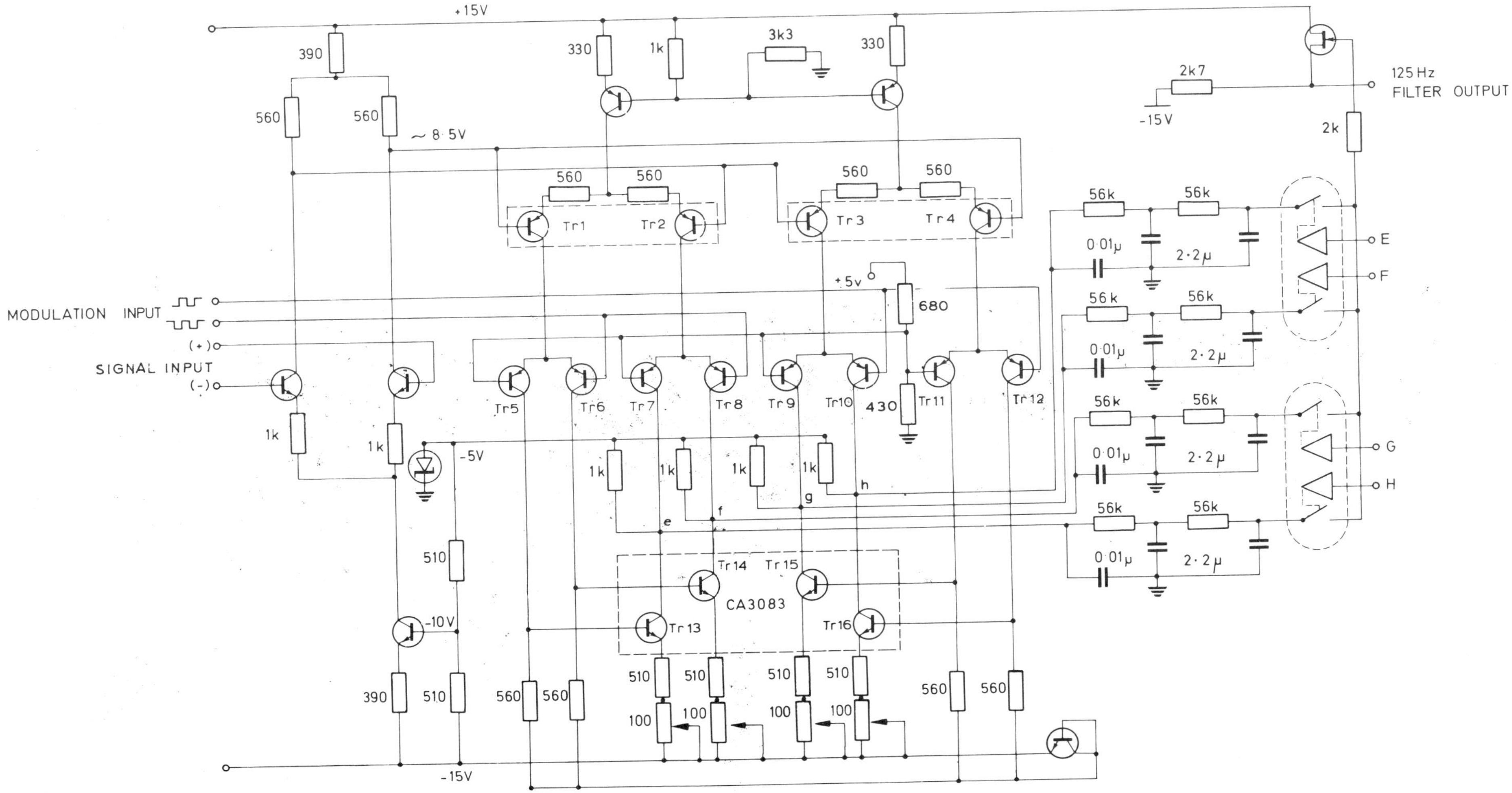

NOTES: PNP = 2N3905 , NPN = 2N3903
MATCHED PAIR , PNP = BFX11 , ANALOG SWITCH DRIVER = DG200

FIG. (5) CURRENT SWITCHED DIGITAL RC FILTER

A REAL-TIME FOURIER ANALYSER FOR THE ANALYSIS OF NON-STATIONARY DATA

J.R. Robertson

Introduction

This paper describes the development of a portable Fourier analyser specifically to cope with the requirements of narrow-band real-time spectral density analysis in the solid-fuel rocket-motor field. The frequencies of interest lie in the band (0-10) KHz and time histories of any static firing are characterised by a variable length output with rise and fall times of a few milliseconds. During the period of useful output, the thrust may vary due to design considerations or to imperfections in the manufacturing process. In addition, the ignited motor has a time-varying mass and may be acoustically enshrouded due to the protective shelter. Hence the parameters of the firing will have definite periods when the related data is non-stationary and possibly appearing as random bursts.

Often a transient burst is shorter than one second and since an insufficient ensemble of trials cannot be mounted immediately or if they could be guarantee the transient to occur over the same time period, all performance data must be extracted from a single trial. Therefore traditional attempts at estimating the power spectrum of statistical data by modifying the time history and smoothing over many periods of time can be irrelevant, since sufficient degrees of freedom may not be available. Conversely in a very long burning motor, there maybe periods of randomness which can be treated as stationary and the bandwidth/time product is sufficiently large to draw conclusions from the estimates.

Because the motor is the part of the final product which gives rise to the major vibrational forces which the missile must withstand, the frequency distribution of these forces is a vital parameter for which adequate mechanical damping of the complete structure must be arranged. Accelerometer readings give a valid measure of this performance and these can be related to velocity and displacement spectral content if these parameters are further required. Typically, acceleration spectral-density analysis calls for the examination of sections of the time history to investigate burst of spectral activity, as well as give average readings for the whole firing. Hence the analyser must be able to lock-out all data until commanded to do otherwise.

Limitations of Current Real-Time Analysers

The stimulus for this analyser resulted from there being apparently no commercial instrument available, outside a general-purpose computer, to tackle completely the above problem. The properties and shortcomings of possible instruments will therefore be highlighted.

Current commercial narrow-band real-time spectrum analysers are broadly of two types. They use either time comparison (TC) or fast Fourier transform (FFT) techniques on digitised data to synthesise the equivalent of a plurality of parallel filters over the range in which the data is to be analysed. Either method allows for versatile range

J. R. Robertson is with Imperial Metal Industries Limited.

switching, with subsequent bandwidth variations, without the complications of having to change physically each filter as in the parallel filter approach. However the FFT processor can cost three times more than the TC processor.

The former uses one recirculating data buffer store with the store contents being updated by one fresh sample for every circulation (lap) or multiple thereof. Hence the original sampled data is time compressed by a multiple of m : 1 where m is the total number of synthesised filters. Additionally, if N sample points are required to fill the buffer, then the processing time for real-time throughput is mN x (the buffer clock rate) where $m \not> N/2$. The time compressed signal is passed through a D/A converter, heterodyned through a fast responding filter and the sampled envelope taken as a measure of the Fourier amplitude.

Unfortunately this method of storage, in which the oldest sample is always removed per circulation or multiple thereof, is unsatisfactory for rapidly changing data, since the compressed time history seen by each of the m filters is changing at the buffer clock rate. This is most important during the first segment to be analysed, since the data passed through the first filter will have disappeared by the time the mth filter is brought into use. Further, if a number of segments are averaged, each filter will have viewed the complete history minus the number of samples removed before that filter first viewed the data. This could be crucial if there was an abundance of high frequency components in the first segment.

The latter technique uses a discrete mechanisation of the Fourier transform to obtain the magnitude of the spectral components. Two buffer stores are used, with read/write access to each word, the contents of one buffer being analysed whilst the other is filled with the next consecutive series of data samples. Effectively the stored data is time compressed since the processing is performed according to the timing cycle of the processor's arithmetic unit and read/write cycle. Additionally the processor makes use of redundancy in certain of the (N/2 x N) complex products such that the analysis time is governed by a logarithmic law. The time to process N/2 filters would be $N/2 \log_2 N$x (cycle time for one complex product) if the real and imaginary parts can be processed concurrently in a complex arithmetic unit (1). Additionally the N/2 filters view the same data segment. When

$$\frac{N}{\log_2 N} > \frac{\text{FFT cycle time}}{\text{TC clock rate}}$$ the FFT processor will be the faster.

For a time ratio of 10 : 1 this would occur for $N \doteqdot 64$ and this processing speed advantage can be used to give either more finely resolved spectra, in the same processing time as a TC model, or allow for the real-time analysis of higher frequencies for the same sample size. This enhanced processing speed is obtained at the joint cost of more complex buffer stores and either additional hard-wired logic or software and memory to control the ordered though irregular manipulation of complex products. Additionally, with a software controller, the FFT processor may of course calculate other statistical parameters as well.

The speed-up process of each has been explained and for the examination of non-stationary data two buffer stores are obviously required. However there is a further advantage to the FFT processor concerning the conversion of the amplitude spectra to its power equivalent. Since this is performed digitally, a true square output will be obtained which maintains the dB resolution of the original A/D converter. Since the TC processor is purely analogue, only 30 dB of range would be obtained for a true power output. When this performance is inadequate, a 40 dB range can be obtained by linearly detecting and sampling the output from each filter. This method produces a pseudo r.m.s. output which has a known relationship to the measured absolute average, if the statistics of the signal are known. With non-stationary data this approach is obviously unwise.

Finally, both TC and FFT machines average the calculated spectra of each segment digitally and this gives rise to a further problem in analysing finite data. Typically, the anti-alias filters, which determine the length of the data segment, have ranges which are limited to increments of 1, 2 and 5. Accordingly, the data sampling rate and the N samples per store define the segmented time history inversely proportional to the 1, 2 and 5 progression of range frequencies. Further, unless the short term spectra are averaged by a general-purpose computer, the attendant ensemble averager will be in values of 2^a. Consequently between a and a-1, there will exist $2^{a-1}-1$ values which cannot be used and therefore close tolerance averaging is just not possible, unless the trial duration and 2^ax(segment length) coincide. For example, if the average is required over a 100 seconds and the chosen range implies 1 second segments, 64 or 128 may be dialled on the averager. The former is insufficient and the latter biases the average reading with unwanted trailing zeros in the time history.

Proposed Instrument

An instrument is proposed with two buffer stores operating in an alternating write/recirculate mode and the data analysed by linearly time compressing it, since the maximum range of interest will be quite adequately accommodated by consecutive serial computation. The time-compressed data, in conjunction with a read-only-memory (ROM)sine/cosine look-up table, is used to generate a discrete Fourier transform of each data segment. Further ROMS allow the square magnitude of the Fourier components to be obtained, the phase terms being disregarded for the present application. All averaging is performed in the accepted manner of powers of two, to be followed by a log ROM output to a D/A converter which preserves the dB variation of the input A/D conversion. This arrangement in itself however does not get over the requirements for close tolerance averaging, unless more freedom is given in the selection of the frequency range. Additional freedom is made possible by using a novel hybrid anti-alias filter in which digital control of the cut-off frequency is very easily obtained from a thumbwheel switch. Although the switch is arranged to be four decades in length, giving 1Hz resolution, any number could have been used to give fractional Hz resolution. In combination with the binary power averager, an accurate average is now possible over any length of data with the proviso that you must accept the frequency resolution involved. A prior knowledge of a trial length is obviously required but this can be easily obtained if the data is first recorded. The instrument can be more fully examined by detailing its four sections in conjunction with the block diagram.

Data Reduction

This section changes the analogue input to a suitable digital equivalent for subsequent processing. The input amplifier/attenuator arrangement gives +50 to -30dB variation in 6dB steps and scaling of the data can be viewed on the built-in oscilloscope, with a front-panel optical indication if the A/D converter is overloaded in either direction. Low-pass "integrating" sections, valid from .2Hz, are available if velocity and displacement spectra are required.

Due to the multiplicating effect of the sampling process, anti-alias filters are required to remove possible folded spectral components. These filters are high-order low-pass sections which, in this case, are implemented by using a sequence of first-order capacitor-ratio commutated-filter[(2)] networks. The time constant of such a network is related to the ratio between two capacitors and the commutating rate. This facility allows the frequency range to have a variable upper value of (10-9,999)Hz in steps of 1Hz and the corresponding data segments to be varied from $(10^{-4} \rightarrow 10^{-1})Ns^{-1}$ seconds, where N is the number of samples per segment and s the normalised sampling rate relative to the upper range frequency. The frequency range is controlled by a four-decade thumbwheel switch. For .1% resolution throughout the whole range, six decades would be required. One property of a commutating filter is that further aliases may be introduced and a boundary filter is used to preclude them. These aliases can be arranged to occur sufficiently far away from the cut-off point that the boundary filter is of simple design and needs only to be related to the most significant decade of the range control.

The A/D converter has an 8 BIT + SIGN output in two's complement code. This is converted to magnitude and sign to be compatible with the ROM sine/cosine outputs and amplitude trigger facility. Note that "windowing" of the data has not been introduced since amplitude distortion of non-stationary data would occur.

Fourier Processor

The two recirculating memories are made from MOS dynamic shift registers of 1024 BITS length. Each one is alternately clocked with WRITE/RECIRC. pulses and the time compressed data fed through a complex multiplier. Each one of the two multipliers is fed with either a discrete sine or cosine function from a composite ROM with an 8 BIT accuracy. Since each ROM segment has only a 5 BIT address, four are paralleled to resolve 90° into 128 discrete weights, such that this composite ROM allows a maximum of 256 synthesised parallel filters.

The Fourier components, $F(k) = \sum_{n=0}^{N-1} x(n)\exp(-j2\pi knN^{-1})$, are calculated in a serial manner by loading k into a summing device which is incremented by k for every value of n. The resultant output is used to address both types of ROM, without seeking recourse to the computational redundancy of the FFT technique. Only the 8 significant BITS next to the sign value are used to form the squared magnitude of F(k) from the separate quadrature components $F_s(k)$ and $F_c(k)$. In this case phase components are of no interest. The squaring is performed digitally by further ROMS which retain both the 48 dB range of the input and keep the data in its true form, without making assumptions about the probability distribution of the time history.

Ensemble Averager and Display

A static shift register, 256 BITS long, is used to gather data at a rate commensurate with the processing. By recirculating the data and adding, an approximate running average mode can be synthesised if the output is divided by 2^a as 'a' travels from 0 to the value selected. The output is in a 16 BIT format and it is impractical to use a 16 BIT D/A converter to drive the display. It is therefore expedient to convert the power value to its logarithmic equivalent. Logarithms to the base two are taken instead of ten since only a constant scaling factor is involved. Simple decoding of the position of the most significant bit of the 16 BIT format gives the characteristic of the logarithm as a 4 BIT number. The next four significant bits are the address for a ROM whose output is the equivalent 4 BIT mantissa correct to $\pm$ .1dB. This composite 8 BIT number can be used to drive a simple 8 BIT D/A converter and therefore preserve the accuracy of the original conversion.

Digital Controller

The range switch provides a proportional clock, derived from a crystal oscillator, to the commutated filter, the sample and hold and A/D converter. Both the sampling rate and the crystal clock are used to write and recirculate the data in each memory alternately. There is also a free running 10 Hz clock which can be used, in conjunction with a four-decade counter, to provide a further trigger depending on the time history elapsed before the analysis begins. Alternatively, a clock recorded with the data may be used. Analysis can be stopped when the selected number of averages is reached.

The input data rate to each D/A converter, which jointly control the display, can be controlled by the lap marker in the spectrum mode, the write rate in the time mode, or by a variable clock rate when the analysis is finished. The output display may be either an oscilloscope or plotter.

Acknowledgement

This paper reports work carried out with the support of the Procurement Executive, Ministry of Defence.

References

(1) Klahn R et al : "The time-saver : FFT hardware" Electronics, June 24, 1968 pp (92 - 97)

(2) Robertson J. R. : "A capacitor-ratio commutated-filter". To be published.

FOOTNOTE: This paper is subject to a patent application.

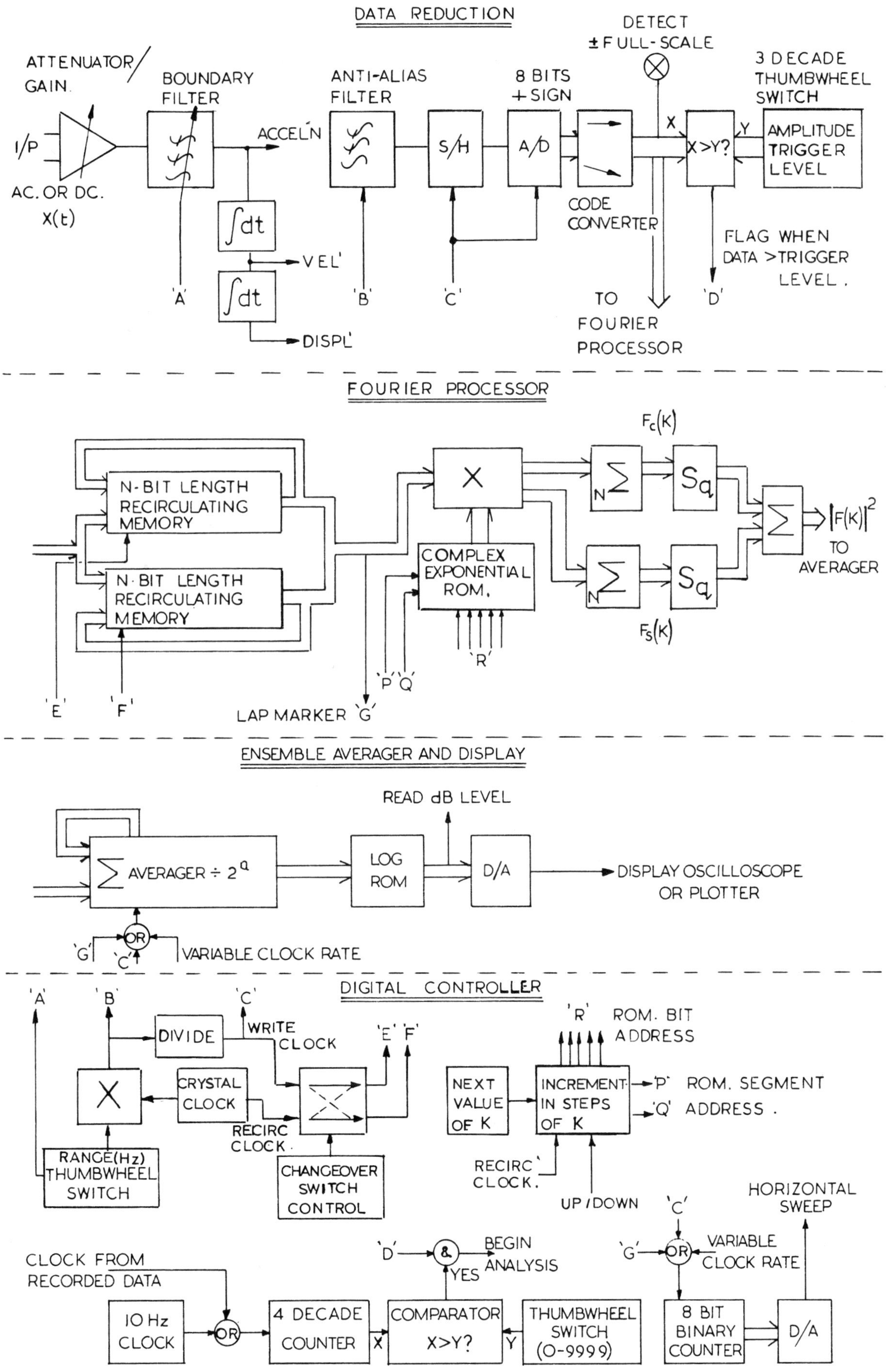

BLOCK DIAGRAM OF ANALYSER

RECURSIVE ALGORITHMS FOR DIGITAL TRANSFORMATIONS

T. S. Durrani and J. M. Nightingale

Introduction

The power spectral density function is an important parameter of many stochastic distributions encountered in signals associated with physical systems. In principle, it accomplishes a mapping of the properties of stochastic data from the time to the frequency domain. This mapping usually involves a set of orthogonal transformations.

Much attention has been devoted to the problem of efficiently computing these transformations and hardware and software variants of the fast Fourier transform are much in favour. The present work is concerned with different and more fundamental aspects of spectrum estimation of time series.

Empirical spectral analysis of digital data gives a statistical estimate of what is considered as a significant probability function of the underlying stochastic process. Such estimates should be statistically reliable. At the simplest level this involves the choice of some basic parameters such as sampling rates, data lengths, frequency resolution of analysis. Hence, current approaches to spectrum estimation depend upon the availability of suitably long records of the process for adjusting these parameters before computations can commence. This generally involves excessive storage and does not necessarily lead to flexibility in analysis.

Recursive estimation is an efficient technique for estimating parameters of random data according to a minimum error variance criterion[1]. Though the theory of least mean square errors is implicit in devising a spectral estimator, the advantages of an arbitrary choice of estimating parameters offered by recursive methods have prompted this presentation. In this context regularity in size and simplicity of format of recursive estimators can be exploited to establish compact and modular designs for hardware realisations of the underlying algorithms.

Recursive Algorithms

Considering a stochastic data sequence $\{X_k\}$ for a normalised Nyquist frequency of ½Hz, a recursive estimator for the power spectral density function, with a frequency resolution of 1/N Hz, may be derived as[2]

$$\underline{\Phi}(m+1) = \underline{\Phi}(m) + \frac{1}{m}\left[X_m(\underline{H}X_m + 2\mathrm{Re}\underline{B}(m)) - \underline{\Phi}(m)\right] \quad (1)$$

$$\underline{B}(m+1) = (\underline{B}(m) + \underline{H}X_m)\,\underline{W} \quad (2)$$

The authors are with the Control Group, University of Southampton, Southampton, SO9 5NH.

with conditions $\underline{\Phi}(0) = \underline{0}$; $\underline{B}(qN) = \underline{0}$, $q = 0,1,2 \ldots$ (3)

where

$$\underline{\Phi}^T(m) = \left[\Phi(0)\ \Phi(\tfrac{2\pi}{N})\ \Phi(\tfrac{4\pi}{N})\ \ldots\ \Phi(\tfrac{\pi}{2})\right]_m \tag{4a}$$

$$\underline{B}(m) = \left[B_0\ B_1\ \ldots\ B_2\ \ldots\ B_{\frac{N}{2}}\right]_m \tag{4b}$$

$$\underline{W}^T = \left[1\ \ e^{-\frac{j2\pi}{N}}\ \ e^{-\frac{j4\pi}{N}}\ \ldots\ e^{-\frac{j\pi}{2}}\right] \quad (\tfrac{N}{2}+1) \times 1 \text{ unit vector} \tag{4c}$$

$\underline{\Phi}(m)$ is a real vector containing instantaneous values of the spectral ordinates of the observed process $\{X_k\}$. $\underline{B}(m)$ is a complex dummy vector which contains instantaneous (transient) values of the discrete Fourier transform of the data sequence up to the current observation X_m. $\underline{W}$ is the trigonometric coefficient vector.

The above algorithm may be recast into a state space notation as

$$\underline{Z}(m+1) = \underline{A}(m+1,m)\ \underline{Z}(m) + \underline{\Gamma}(m+1,m)\ \underline{H}_1 X_m \tag{5}$$

where

$$\underline{Z}^T(m) = \left[\underline{\Phi}^T(m)\quad \underline{B}(m)\quad \underline{B}^*(m)\right] \tag{6}$$

$$\underline{\Gamma}^T(m+1,m) = \left[\frac{X_m}{m+1}\underline{I}\quad \underline{G}\quad \underline{G}^*\right] \tag{7}$$

$$\underline{G} = \text{diag}\ \underline{W} \qquad (\tfrac{N}{2}+1) \times (\tfrac{N}{2}+1) \text{ diagonal matrix} \tag{8}$$

$$\underline{H}_1^T = \left[\underline{H}^T\quad \underline{H}^T\quad \underline{H}^T\right]$$

$$\underline{A}(m+1,m) = \begin{bmatrix} \frac{m}{m+1}\underline{I} & \frac{X_m}{m+1}\underline{I} & \frac{X_m}{m+1}\underline{I} \\ \underline{0} & \underline{G} & \underline{0} \\ \underline{0} & \underline{0} & \underline{G}^* \end{bmatrix} \tag{9}$$

* denotes complex conjugate.

Notice that both $\underline{\Gamma}(m+1,m)$ and $\underline{A}(m+1,m)$ are excessively sparse matrices.

Considerable simplification is achieved if a normalised power spectrum is estimated, i.e.

$$\underline{S}(m) = \underline{\Phi}(m)/\sigma_m^2 \tag{10}$$

where σ_m^2 is a recursive estimator for the variance of the process $\{X_k\}$, so that for a zero mean $\{X_k\}$,

$$\sigma_{m+1}^2 = \sigma_m^2 + \frac{1}{m+1}(X_m^2 - \sigma_m^2) \tag{11}$$

The above relationships then reduce to

$$\underline{Q}(m+1) = \underline{Q}(m) + 2\mathrm{Re}\,\underline{B}(m)\,X_m \tag{12}$$

$$\underline{B}(m+1) = (\underline{B}(m) + \underline{H}\,X_m)\,\underline{W} \tag{13}$$

$$\alpha_{m+1} = \alpha_m + X_m^2 \tag{14}$$

with

$$\underline{B}(qN) = \underline{0} \qquad q = 0,1,2 \ldots$$

where

$$\underline{S}(m) = \underline{H} + \underline{Q}(m)/\alpha_m \tag{15}$$

A closed form solution may be evaluated for Eq. (1), (2) or (12-14). For Eq. (1) this yields any pth element (spectral ordinate) of the vector $\underline{\Phi}(m)$ as

$$\Phi\left(\frac{\pi p}{N}, m\right) = \frac{1}{m}\left[\sum_{u=1}^{q}\left(\sum_{k,s=(u-1)N}^{uN-1}\sum X_k X_s\, e^{-\frac{j2\pi p(k-s)}{N}}\right) + \sum_{k,s=qN}^{m-1}\sum X_k X_s\, e^{-\frac{j2\pi p(k-s)}{N}}\right] \tag{16}$$

for $p = 0,1,2 \ldots \frac{N}{2}$; $qN < m < (\overline{q+1})N$.

The above estimate of the spectral ordinate converges in the mean to

$$\underset{m\to\infty}{\mathrm{Lt}}\; E\left|\Phi\left(\frac{\pi p}{N}, m\right)\right] = \sum_{k=1-N}^{N-1}\left(1 - \frac{|k|}{N}\right) R(k)\, e^{-\frac{j2\pi pk}{N}} \tag{17}$$

while

$$\underset{m\to\infty}{\mathrm{Lt}}\; \mathrm{Var}\left[\Phi\left(\frac{\pi p}{N}, m\right)\right] \to 0 \tag{18}$$

where $R(k)$ is the autocorrelation function of the process $\{X_k\}$ for lag k.

Cross-spectrum Estimator

For two data sequences $\{X_k\}$, $\{Y_k\}$, a recursive estimator may be derived for estimating their cross-spectra. Under conditions of zero mean and normality, and stationarity

$$\underline{\Psi}_{XY}(m+1) = \underline{\Psi}_{XY}(m) + \frac{1}{m+1}\left[X_m\,\underline{C}(m) + Y_m\,\underline{B}(m) + \underline{H}X_m\,Y_m - \Psi_{XY}(m)\right] \quad (19)$$

$$\underline{B}(m+1) = (\underline{B}(m) + \underline{H}X_m)\,\underline{W} \quad (20)$$

$$\underline{C}(m+1) = (\underline{C}(m) + \underline{H}Y_m)\,\underline{W}^* \quad (21)$$

with conditions $\underline{\Psi}_{XY}(0) = \underline{0}$; $\underline{C}(qN) = \underline{B}(qN) = 0$, $q = 0,1,2 \ldots$

where $\underline{\Psi}_{XY}(m)$ is a $(N/2 + 1) \times 1$ complex vector containing instantaneous values of the cross-spectral density ordinates established at a (normalised) frequency resolution of 1/N Hz. The real and imaginary parts of the elements are the co- and quadrature components of the cross-spectral ordinates. $\underline{B}(m)$ and $\underline{C}(m)$ contain the instantaneous values of the discrete Fourier transforms of the stochastic sequences $\{X_k\}$, $\{Y_k\}$ respectively.

The convergence of the estimator is apparent, if we consider any pth element of the complex vector $\underline{\Psi}_{XY}(m)$

$$E\left[\Psi(\tfrac{\pi p}{N}, m)\right] = \sum_{k=1-N}^{N-1} (1 - \tfrac{|k|}{N})\, R_{XY}(k)\, e^{-\frac{j2\pi(k)}{N}}$$

for large m, and

$$\operatorname*{Lt}_{m\to\infty} \operatorname{Var}\left[\Psi(\tfrac{\pi p}{N}, m)\right] \to 0$$

where $R_{XY}(k)$ is the cross-correlation function between the two sequences $\{X_k\}$, $\{Y_k\}$.

The recursive algorithms posed here establish a procedure for utilising each new observation of a data sequence to refine the previous spectral estimate. The latest observation is appropriately weighted to determine its contributions to spectral ordinates at various frequencies. These contributions are then combined with the previous spectrum values to yield an updated spectrum.

The sequential estimation procedure can be terminated at any stage of the iterations. The result obtained on termination is the optimal spectrum estimate based upon all the accumulated data up to the instant of termination. An interesting feature is that at each iteration the computations are fixed in size and format, and there is no need for maintaining a complete record of all the previous observations. Frequency resolution can be varied with ease by refreshing the $\underline{W}$ vector with new sets of trigonometric coefficients.

Aspects of Smoothing

Conventional techniques for data/frequency smoothing to minimise leakage effects in digital spectra are well established[3-5]. Prior to proposing a smoothing scheme it is useful to determine the spectral window associated with the recursive estimators of the previous sections.

From Eq. (16), for $m = qN+r$, $q = 0,1,2 \ldots$; $r = 0,1,2 \ldots (N-1)$,

$$E\left[\Phi\left(\frac{2\pi p}{N}, m\right)\right] = \frac{1}{m}\left[qN \sum_{k=1-N}^{N-1} \left(1 - \frac{|k|}{N}\right) R(k)\, e^{-\frac{j2\pi pk}{N}} + r \sum_{k=1-r}^{r-1} \left(1 - \frac{|k|}{r}\right) R(k)\, e^{-\frac{j2\pi pk}{N}} \right] \qquad (22)$$

For a Nyquist frequency of F/2, the above equation may be transformed to yield the expected value of the spectral ordinate at any frequency f as

$$E\left|\Phi(f,m)\right| = \int_{-F/2}^{F/2} \Phi_T(f-g) \left\{\frac{qN}{m} \frac{\sin^2 \frac{\pi g N}{F}}{N \sin^2 \frac{\pi g}{F}} + \frac{r}{m} \frac{\sin^2 \frac{\pi g r}{F}}{r \sin^2 \frac{\pi g}{F}}\right\} dg/F \qquad (23)$$

for

$$-F/2 \leqslant f \leqslant F/2$$

where $\Phi_T(f)$ is the true value of the spectral density function at f Hz. Thus the spectral window associated with the recursive estimators is composed of two windows. The main window is given by

$$D_1(f) = \frac{qN}{m} \frac{\sin^2\left(\frac{\pi f N}{F}\right)}{N \sin^2 \frac{\pi f}{F}}$$

which is a Fejer kernel with major lobe height of N(qN/m), and zeros at intervals (F/N) Hz. Superimposed on this is a smaller window

$$D_2(f) = \frac{r}{m} \frac{\sin^2 \frac{\pi f r}{F}}{N \sin^2 \frac{\pi f}{F}}$$

with a major lobe of height r(r/m) and zeros at (F/r) Hz. It has much broader (and transient) side-lobes which introduce a ripple effect on the main window, and hence exaggerate spectrum leakage from the main window.

From Eq. (23) it is obvious that $\Phi(\frac{\pi p}{r}, m)$ conforms to the conventional estimator of the spectral density function, since

$$\underset{m\to\infty}{\mathrm{Lt}}\; E\left[\Phi(f,m)\right] = \int_{-f/2}^{F/2} \Phi_r(f-g) \frac{\sin^2 \frac{\pi g N}{F}}{N \sin^2 \frac{\pi g}{F}} dg/F \quad \forall f \qquad (24)$$

To combat the effect of the smaller cyclic spectral window within the main spectral window, a cyclic data smoothing procedure may be employed such that the input to the recursive algorithms is the smoothed sequence $\hat{X}_k = d_k X_k$, where the data window $\{d_k\}$ is such that $\{d_k = d_{k+qN}$, $q=0,1,2 \ldots\}$. Further details on data windows may be found in Ref. 4. A data window found to be of special significance in minimising leakage without undue loss of statistical stability

$$d_k = 1 - \left| \frac{k - \frac{N-1}{2}}{\frac{N+1}{2}} \right|^5$$

System Organisation

The system realisation hinges on the design of four functional units which would form the basic structure for implementing the algorithms in hardware. Following current trends in computer architecture this would be organised into an arithmetic unit, memory unit, control unit and input-output units. In principle, it is the design of the control unit and its associated timing sequences which is considerably simplified. Storage requirements are dictated only by the length of the vector $\underline{Z}(m)$ in Eq. (5), and not the overall data length. As the vector $\underline{W}$ is time invariant, it could be directly accessed from a ROM containing trigonometric coefficients.

The choice of word-length is influenced by underflow and overflow requirements and round-off noise generation. Upper bounds have been derived for recursive algorithms to establish scaling factors and normalising constants. For a data sequence with $\max|X_m| < 1$, any element of the vectors $\underline{Q}(m)$, Eq. (12), or $\underline{S}(m)$, Eq. (15) have the following bounds (for $qN < m \leq (q+1)N$)

$$\left[Q(\tfrac{\pi p}{N}, m)\right] \leqslant q(N^2 - N) + (m - qN - 1)(m - qN)$$

and

$$\left[S(\tfrac{\pi p}{N}, m)\right] \leqslant \frac{qN^2 + (m-qN)^2}{mN} \; ; \text{ i.e. } \mathrm{Max}\left|S(\tfrac{\pi p}{N}, m)\right| < 1 \quad \forall p$$

Thus estimation of normalised power spectral ordinates would always avoid overflow.

Round-off noise generated at each stage of the algorithms may be determined to provide upper bounds on the accuracy of the computations and thus indicate acceptable word-lengths. Using a 'forward error analysis' approach this bound may be shown to be

$$\text{R.M.S.}\left(\frac{\text{Noise}}{\text{Signal}}\right)_{O/P} \leqslant \frac{2}{3}\sqrt{2N}\,\sigma_e$$

where σ_e^2 is the variance of the round-off variables uniformly distributed in the interval $(-2^{-t}, +2^{-t})$, t being the word-length of the accumulators.

References

1. Kalman, R.E. : New methods and results in linear prediction and filtering theory. Tech. Rept. 61-1, RIAS, Baltimore, Md, 1960.
2. Durrani, T.S. : Ph.D. Thesis, University of Southampton, U.K. 1970.
3. Cooley, J.W., Lewis, P.A.W., and Welch, P.D. : The Fast Fourier Transform Algorithm and its Applications. IBM Res. paper RC-1743, 1967.
4. Durrani, T.S., and Nightingale, J.M. : Data windows for digital spectral analysis. Proc. I.E.E., 1972, V. 119, pp. 343-352.
5. Sloane, E.A. : Comparison of linearly and quadratically modified spectral estimates of Gaussian signals. I.E.E.E. Trans. 1969, Au-17, pp. 133-137.

ACCURACY OF TRANSFER FUNCTION ESTIMATES DERIVED FROM NUMERICAL FOURIER TRANSFORMS

M.T.G. Hughes and P.E. Wellstead

1. Introduction

Input/output relationships in linear dynamic systems can be represented in a variety of ways, but the steady-state frequency response function is probably the representation most widely used in experimental studies. The conventional transfer function analyzer (TFA) injects a sinusoidal test perturbation at the input of a system under test, and employs a two-channel correlator to estimate the in-phase and quadrature components of the complex system gain at the test signal frequency.

This type of TFA has found wide acceptance, but suffers from two main limitations:

(i) Its use is restricted to applications in which sinusoidal test perturbations are acceptable.

(ii) Since one such test must be carried out for every frequency which is of interest, the number of tests required is often large, and the overall test procedure tends to be relatively time-consuming.

These limitations may be effectively avoided by using techniques of transfer function estimation based on numerical Fourier transformations of input/output data chosen to span a suitable range of frequencies. Special types of digital instrumentation, often employing highly efficient algorithms for "fast Fourier transformation" have been extensively developed for such purposes in recent years, and their cost/performance characteristics have followed similar favourable trends to those of other devices based on digital electronics. Such systems may therefore be expected to become increasingly available in applications where transfer function analysis must be carried out under difficult environmental conditions.

Whilst the availability of such aids to dynamic analysis undoubtedly has many attractions it is also most important that the user should understand the fundamental limitations and sources of errors in the relatively sophisticated techniques employed. The aim of this paper is to discuss the main sources of errors which are inherent in processes of transfer function analysis based on numerical analysis of frequency spectra, and to indicate quantitatively how such errors may be influenced by the conditions prevailing during particular experiments.

2. Computational Procedures

Computational procedures for transfer function analysis may differ appreciably in their details; but in many fundamental respects the factors which limit the accuracy and resolution of almost any particular procedure may be related to corresponding factors in the so-called 'direct' method of spectrum analysis,[1] which is discussed here.

M.T.G. Hughes is with the Inter-University Institute of Engineering Control, University of Warwick.

P.E. Wellstead is with the Control Systems Centre, University of Manchester Institute of Science and Technology.

Referring to Fig. 1, we suppose that the aim is to estimate the complex frequency response function $H(\omega)$ of the linear and time-invariant system shown, by analyzing records of the noise corrupted signals $x(t)$ and $y(t)$ at the input and output of the system respectively. Using the direct method, this may be accomplished as follows:

1. Sample pairs of the signals $x(t)$ and $y(t)$ are digitized sequentially at uniformly spaced time instants (at say Δ - seconds intervals) and stored in blocks or 'frames' of N sample pairs each, as indicated in Fig. 2.

2. As each frame is completed, the samples contained therein are transformed numerically to yield a set of discrete Fourier coefficients for the x and y signals respectively. Thus, if $x_i(r)$, $y_i(r)$ represent the values of x and y in the r'th position of the i'th frame, then the corresponding sets of complex Fourier coefficients for the frame will be given by

$$X_i(k) = \frac{1}{N}\sum_{r=0}^{N-1} x_i(r)\ell(r)\exp\left(\frac{-j2\pi kr}{N}\right)$$

$$Y_i(k) = \frac{1}{N}\sum_{r=0}^{N-1} y_i(r)\ell(r)\exp\left(\frac{-j2\pi kr}{N}\right) \qquad \ldots(1)$$

$$k=0,1,2,\ldots,N-1 \rightarrow r=0$$

In these expressions, the index k specifies the harmonic frequencies of the Fourier series as multiples of the fundamental frequency $2\pi/N\Delta$, and the quantity $\ell(r)$ represents an optional 'data window', or weighting sequence, which is sometimes used to control certain deterministic errors (or bias) in the estimates. More will be said about this presently.

3. The Fourier transform sequences in Eq. (1) are used to calculate further sequences (sometimes referred to as periodograms) defined (again for the i'th frame) by

$$P_x(\omega_k)_i = X_i(k), X_i^*(k),$$

$$P_{xy}(\omega_k)_i = Y_i(k), X_i^*(k) \qquad \ldots(2)$$

$$k=0,1,2,\ldots,N-1$$

Here, the asterisk denotes complex conjugation (changing the sign of the imaginary component) and $\omega_k = 2\pi k/N\Delta$

We may note in passing that, although these quantities are defined over the range $k=0,1,2,\ldots,N-1$, we need only evaluate them in practice over about a half of this range. This is because P_x and the real part of P_{xy} are symmetrical about the mid-point of the indicated range of k and similarly the imaginary part of P_{xy} is skew-symmetrical about this mid-point.

The quantities defined in Eq. (2) are related respectively to the theoretical power spectral density function $S_x(\omega)$ of the signal $x(t)$ and $y(t)$, The periodograms are in fact generally poor estimates of the idealized spectral functions, since they are subject to large random errors and also to deterministic (bias) errors. The bias can usually be made acceptably small by suitable selection of the sampling interval, frame duration, and data-window function $\ell(r)$; and the random errors can be made acceptably small by averaging the computed periodograms over a sufficiently large number of frames, thus:

4. If $\hat{S}_x(\omega_k)$ and $\hat{S}_{xy}(\omega_k)$ are the final estimates of $S_x(\omega)$ and $S_{xy}(\omega)$ respectively, we define

$$\hat{S}_x(\omega_k) = \frac{1}{M}\sum_{i=1}^{M} P_x(\omega_k)_i$$
$$\hat{S}_{xy}(\omega_k) = \frac{1}{M}\sum_{i=1}^{M} P_{xy}(\omega_k)_i \qquad \ldots(3)$$

The sequence of steps 1 to 4 is indicated schematically in Fig. 3.

5. Once the spectral estimates in Eq.(3) have been calculated with acceptable stability and resolution, it is usual to estimate the complex frequency - response function $H(\omega)$ by using the estimator

$$\hat{H}(\omega) = \hat{S}_{xy}(\omega_k) / \hat{S}_x(\omega_k) \qquad \ldots(4)$$

3. Analysis of Errors

Given ideal conditions (implying virtually unlimited observation time, an absence of input measurement noise $m(t)$, and an absence of significant correlation between the input signal $v(t)$ and the output error processes $d(t)$ and $n(t)$, the estimate $\hat{H}$ given by equation (4) can be made to approximate as closely as desired to the true system characteristic H. Failure to satisfy these ideal requirements results in errors, both random and deterministic, and the aim in the following section is to identify and quantify the relevant considerations as simply as possible. This is most conveniently done by considering the effects of the various factors on the bias (mean error) and variance (mean square deviation from the mean) characteristics of the resulting parameter estimates.

3.1 Effects of Noise

The quantities $d(t)$, $m(t)$ and $n(t)$ shown in Fig 1 represent the aggregated effects of numerous complicated phenomena and may be most conveniently treated as samples of stationary random noise. The component $d(t)$ accounts for system responses to extraneous and unobserved inputs, whilst the components $m(t)$ and $n(t)$ arise as a result of errors in measurement and computation processes.

All three of the indicated noise components introduce random errors into the transfer function estimates, and additionally the input measurement noise $m(t)$ introduces a bias error which always tends to deflate the magnitudes of the estimates. If we define

$$B_m(\omega) = [\hat{H}(\omega)]_{mean} - H(\omega) \qquad \ldots(5)$$

to represent the bias associated with $m(t)$, we may readily evaluate this effect as a fraction of the estimated quantity:

$$\frac{B_m(\omega)}{H(\omega)} = \frac{-S_m(\omega)}{S_v(\omega)+S_m(\omega)} \qquad \ldots (6)$$

Here, as usual, $S_m(\omega)$ and $S_v(\omega)$ represent the power spectral densities of $m(t)$ and $v(t)$ at radian frequency ω. The bias error $B_m(\omega)$ is usually negligible, but it can cause significant errors to arise over frequency bands where $v(t)$ contains little power.

A measure of random errors in $\hat{H}(\omega)$ caused by noise is given by the variance (mean square deviation from the mean) of the magnitude $|\hat{H}(\omega)|$ and the phase $\phi(\omega)$ of the complex estimates. Approximate variance expressions are given by:

$$\frac{\mathrm{Var}\left[|\hat{H}(\omega)|\right]}{|H(\omega)|^2} \simeq \mathrm{Var}[\hat{\phi}(\omega)] \simeq \frac{1}{2M}\left[\frac{1}{K^2_{xy}(\omega)} - 1\right] \quad \ldots(7)$$

Where M is the number of data frames (Fig.2) over which the estimates are averaged, and $K^2_{xy}(\omega)$ is the squared coherency spectrum, defined by

$$K^2_{xy}(\omega) = \frac{|S_{xy}(\omega)|^2}{S_x(\omega)\,S_y(\omega)} \quad \ldots(8)$$

This quantity is often estimated from the data in practice, according to Eq(8), and used as a diagnostic aid. Its theoretical relationship to the various noise components can be shown (5) to be given by

$$K^2_{xy}(\omega) = \left[(1 + \frac{S_m(\omega)}{S_v(\omega)})(1 + \frac{S_n(\omega) + S_d(\omega)}{S_z(\omega)})\right]^{-1} \quad \ldots(9)$$

The principal sources of noise components m(t) and n(t) in Fig.1 are:

(i) Quantization Errors at the Samplers
(ii) Arithmetic round-off during data-windowing
(iii) Arithmetic round-off during Fourier transformation.

When (as is usually the case with currently available real-time systems) the computations are carried out using fixed-point arithmetic, the dominant source of errors is item (iii). Welsh(4) has shown that the level of round-off errors in a fixed point Fast Fourier Transform computation can be expressed most simply as a worst-case variance σ_r^2, given approximately by

$$\sigma_r^2 \simeq c\,2^{n-2b}\;\frac{\text{mean square value of input to the algorithm}}{\text{mean square value of output from the algorithm}} \quad \ldots(10)$$

where c is a constant which takes values between approximately 1 and 10, depending on the detailed arithmetic procedures used, b is the word length employed in the computer,
$n = \log_2(N)$, where N is the frame size of the data sample. It is assumed that N is an integral power of 2, so that the most efficient form of the FFT is employed.

It is clear from Eq(10) that the word length of the computer or FFT machine imposes a limit on the frame size N, beyond which round-off errors become obtrusive.

3.2 Frequency Resolution

Transfer function estimates based on data frames of finite duration will always be subject to bias errors to some extent. The extent of the bias will depend upon the frame duration $N\Delta$ and the resolving power of the data window spectrum $L(\omega)$ (which is the Fourier transform of the data window sequence $\ell(r)$ used in Eq.(1)) in relation to the spectral densities $S_x(\omega)$ and $S_{xy}(\omega)$.

For small bias, the spectrum of $L(\omega)$ should take the form of a narrow pulse when compared with $S_x(\omega)$ and $S_{xy}(\omega)$. Both the magnitudes and the phases of the latter spectra should exhibit only 'small' variations over the bandwidth of $L(\omega)$.

This requires a 'low-pass' type of spectral window with rapid cutoff (such as the Generalized Hanning Window[2]), together with a frame duration, $N\Delta$, sufficiently large to resolve the smallest frequency increment which is of interest.

With many kinds of systems, rapidly changing phase characteristics (corresponding, for example, to resonance characteristics) are satisfactorily resolved when the magnitude characteristics are well resolved. An important exception to this occurs in systems with significant pure time delays. These can give rise to rapid phase variations and hence large bias errors[3]. It is thus advisable to trim out such delays by introducing a corresponding delay in the 'x' channel.

3.3 Frequency Aliasing

It is well known that when a continuous signal is sampled at a uniform rate, the frequency spectrum of the sampled sequence is a periodic function consisting of replicas of the original spectrum spaced along the frequency axis at integer multiples of the sampling frequency. Confusion between overlapping replicas (or aliases) of the continuous signal spectrum can thus give rise to aliasing errors in numerically computed spectra, unless the continuous signal spectrum is band-limited to a region below one-half of the sampling frequency (known as the Nyquist frequency) before sampling.

A study of aliasing errors in numerical transfer function estimation has shown that spuriously large random errors can be caused by aliasing, as well as the better understood bias errors. In particular, in the region of the Nyquist frequency, aliasing tends to produce random errors in gain and phase estimates whose variance is inversely proportional to the real part of the cross spectral density $S_{xy}(\omega)$. Consequently, if (as often happens) the cross spectrum possesses only small in-phase components near the Nyquist frequency, the estimates can exhibit extremely large random fluctuations.

In practice, aliasing is suppressed by passing the signals x(t) and y(t) through low-pass filters prior to sampling; but since real filters can only have a finite cutoff rate, aliasing cannot be totally avoided, and so remains a significant problem.

The effect of non-identical anti-aliasing filters introduces a further source of bias errors, but this effect can usually be controlled by suitable stabilization and matching of the filter characteristics.

4. Closed-loop Systems

It is often necessary to measure the transfer function of a part of a system while it is under closed-loop control. A typical situation is illustrated in Fig. 4, where the assumed requirement is to estimate the forward-path transfer function $H(\omega)$ without breaking the feedback loop.

Unfortunately, the estimator defined by Eq(4) generally yields seriously biased results when applied to this problem. This is because the feedback path introduces correlation between the system input x(t) and the disturbances $d_1(t)$ and $d_2(t)$.

One solution of this problem [5],[6] requires the introduction of an independent reference signal from outside the feedback loop, so that it is correlated with the signal component of x(t), but it is uncorrelated with the disturbances. The reference signal r(t), shown in Fig. 4 satisfies these requirements, and leads to a new transfer-function estimator:-

$$H(\omega) = \hat{S}_{ry}(\omega) \Big/ \hat{S}_{rx}(\omega) \qquad \ldots(11)$$

It has been shown that this estimator is not biased by loop disturbances and that the resulting gain and phase estimates $|\hat{H}(\omega)|$ and $\hat{\phi}(\omega)$ are reasonably well-behaved under practical conditions.

An approximate expression for the variance of the estimates is given by

$$\frac{\mathrm{Var}[|H(\omega)|]}{|H(\omega)|^2} \simeq \mathrm{Var}[\hat{\phi}(\omega)] \simeq \frac{1}{2m}\left[\frac{S_1(\omega)}{S_x(\omega \; |G(\omega)|^2}\right] \quad \ldots(12)$$

where $G(\omega)$ is the closed-loop system transfer function, and $S1(\omega)$ is the power spectral density of disturbance signal $d1(t)$.

This result suggests, interestingly, that to a first order of approximation, the variance of the forward-element estimator is not affected by noise in the feedback path. Smoothing of these estimates may be achieved in the usual way by increasing the number of frames, M.

5. Conclusions

Based upon the results presented in the previous paragraphs we can outline the necessary features of a general transfer function analyzer using numerical Fourier transforms. The principal components of such a machine are indicated in figure 5. There is insufficient space in a presentation of this nature to comment in detail on the machine, instead we list our key points:

1. The figure shows a three-channel analyzer for closed-loop systems; the machine can be used for open-loop systems by the addition of a simple switching arrangement.
2. Analogue anti-aliasing filters are illustrated, however multi-rate digital filters could be used with the advantage that the filter matching problem is removed. Digital filters would however add considerably to the measurement noise.
3. The r(t) and x(t) channels contain variable delay elements- these are used to trim-out large time delays which would otherwise bias the estimates (section 3.1.).
4. The data window elements operate from a precomputed look-up table or special function generator. If the generalised Hanning window is used then the sin/cos generator for the FFT units may be used.
5. Three separate FFT units are shown, these need not be completely independent but could consist (for example) of a common indexing element serving three arithmetic elements.
6. The computational procedure of section 2 was based upon data segmented into consecutive frames. This is unnecessarily restrictive, and the machine should allow the more flexible procedure of overlapping frames (for low sampling rates or limited data records) and permit gaps between data frames when sampling rates are high.

6. References

1. P.D. Welsh. "The Use of Fast Fourier Transforms for the Estimation of Power Spectra", IEEE. Trans on Audio and Electroacoustics, Vol. AU-15, 2, 1967.

2. E.A. Sloane. "Comparison of Linearly and Quadratically Modified Spectral Estimates of Gaussian Signals", IEEE. Trans on Audio and Electroacoustics, Vol. AU-15,2, 1969.

3. M. Akaike. "On the Statistical Estimation of Frequency Response Functions", Ann. Inst. Stat. Maths., Vol. 14, 1, 1962.

4. P.D. Welsh. "A Fixed-Point Fast Fourier Transform Error Analysis", IEEE. Trans on Audio and Electroacoustic, Vol. AU-17, 2, 1969.

5. P.E. Wellstead. "Aspects of Real-Time Spectral Analysis". Ph.D. Thesis. University of Warwick 1970.

6. H. Akaike. "Some Problems in the Application of the Cross-Spectral Method" Advanced Seminar on Spectral Analysis, (Editor: Harris), John Wiley, New York 1969.

7. Acknowledgement

The authors wish to acknowledge support received from the Science Research Council for the research reported in this paper.

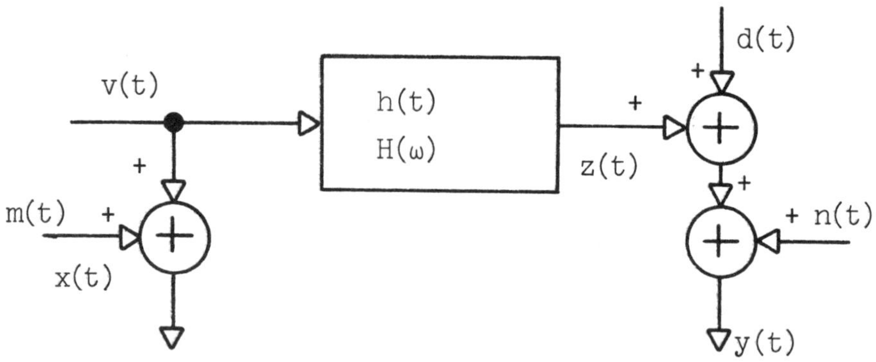

FIG. 1

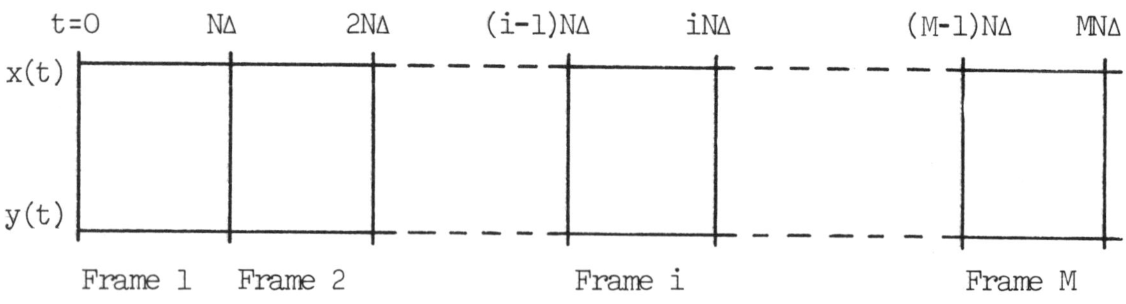

FIG. 2

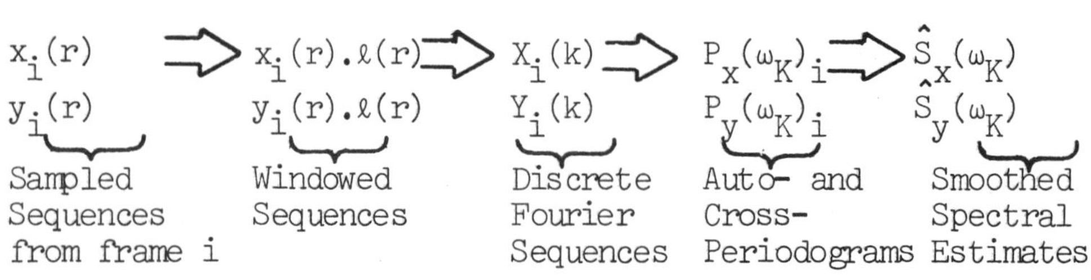

FIG. 3

FIGURE 4 A CLOSED LOOP SYSTEM

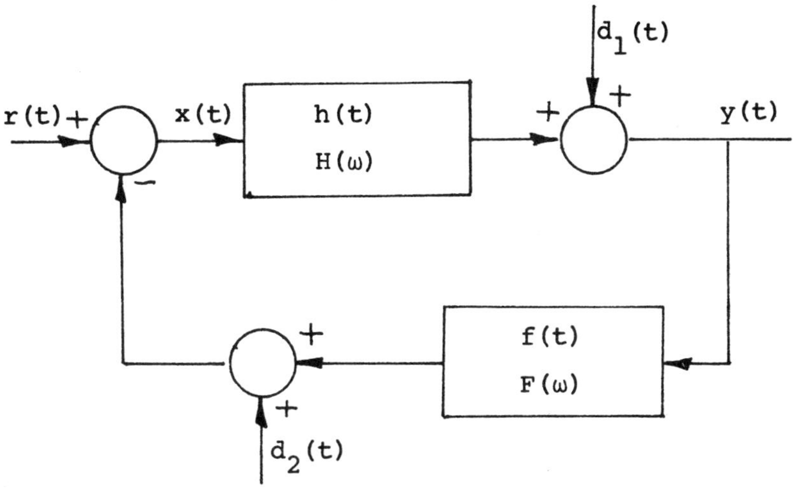

FIGURE 5 A GENERAL DIGITAL TRANSFER FUNCTION ANALYSER

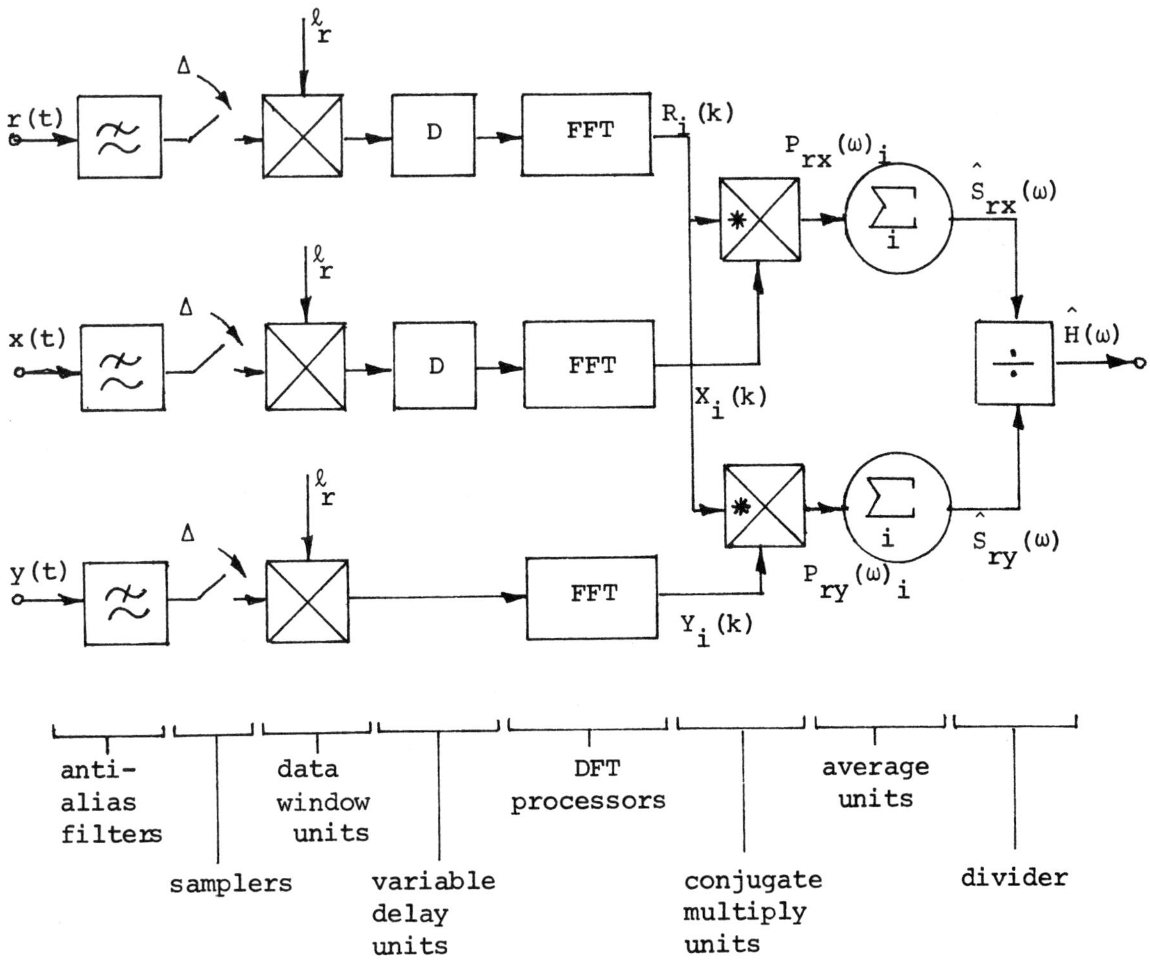

A TRANSFER FUNCTION ANALYZER FOR FEEDBACK SYSTEMS SUBJECT TO TEST SIGNALS OF ARBITRARY FORM

M.T.G. Hughes and J.R. Robertson

1. Introduction

The term 'transfer function analyzer' (or TFA for short) conventionally refers to a device consisting of a two-phase sinusoidal oscillator of variable frequency, and a two-channel correlator. If one phase of the oscillator output is used as an excitation signal to perturb the input of a dynamic system, then the result of correlating the system response with each phase of the oscillator output yields estimates of the real and imaginary components of the complex system gain to sinusoidal inputs at the test frequency. A series of such tests is normally carried out, covering a suitable range of test frequencies, to obtain sets of complex frequency-response loci which convey extensive and useful information on the dynamic system characteristics.

This type of test procedure is widely accepted, and the type of TFA just described has been developed and refined (1). One of the main limitations of such techniques, however, is in the need to subject the system under test to an extended programme of sinusoidal perturbations. Instances do occur where the use of simple periodic test signals may be undesirable, for reasons such as cost, convenience, or even safety, and it was with such considerations in mind that the work described here was undertaken.

The principal aim has been to investigate simple and mobile alternatives to the conventional TFA which will produce accurate transfer-function estimates for systems operating in-situ under typically adverse industrial conditions, without necessarily requiring the use of periodic test signals.

The approach adopted in this paper employs specially devised parameter-tracking feedback loops operating on components of the system input/output signals filtered into a narrow frequency band. On this basis, it is possible to devise a number of TFA systems which appear promising.

In the following sections, we outline the results of experience gained so far with this type of TFA, and indicate some of the design considerations which arise with this type of instrument. The work described here forms part of a more general investigation into new methods of transfer function analysis. One obvious alternative to the present approach is to use one of the currently available facilities for numerical Fourier analysis. Such procedures are, in fact, discussed elsewhere in this conference (2), and serve as a useful reference against which alternative proposals may be assessed.

M.T.G. Hughes is with the Inter-university Institute of Engineering Control, University of Warwick

J.R. Robertson is with Imperial Metal Industries Ltd., Birmingham and was formerly at the University of Warwick.

2. Basic Principles

In its simplest form, the parameter-tracking technique discussed earlier leads to the basic TFA organization shown in Fig. 1. A prototype instrument based on this principle has been constructed and tested, with promising results (3), (4).

The principles of operation of the scheme shown in Fig. 1 are, briefly, as follows :

The system input and output signals x(t) and y(t) are subjected, by means of an accurately matched pair of heterodyne filters, to an effective band-pass filtering process. A double-line notation is used in the Figure to indicate the 'complex' nature of various quantities, such as the filter outputs $X(\omega,t)$ and $Y(\omega,t)$, for example. The 'real' and 'imaginary' parts of the various quantities correspond in their respective phase relationships to the 'cosine' and 'sine' components of the two-phase oscillator signals applied to the heterodyning filters. The indicated operations such as addition, multiplication, conjugation, etc. are supposed to conform to the usual laws of complex algebra.

The 'complex' quantity $\hat{H}(\omega,t)$ shown in Fig. 1 represents a time-varying estimate of the system gain $H(\omega)$ of the system under test, at frequency ω. In the parameter-tracking loop, $\hat{H}(\omega,t)$ multiplies the 'complex' quantity $X(\omega,t)$, which represents those components of x(t) existing in the region of frequency ω. The result of the complex multiplication is compared with the similarly filtered components of the system output y(t), so forming a 'complex' error vector $E(\omega,t)$, such that

$$E(\omega,t) = \hat{H}(\omega,t).X(\omega,t) - Y(\omega,t) \qquad \ldots\ldots (1)$$

The parameter-tracking loop is designed with the aim of minimizing the 'squared modulus' of $E(\omega,t)$ along a path of steepest descent in the $\hat{P}$, $\hat{Q}$, plane. This is achieved by choosing a law of parameter adjustment which satisfies

$$\frac{d\hat{P}(\omega,t)}{dt} = -K_1 \frac{\partial}{\partial\hat{P}}(|E(\omega,t)|^2)$$

$$\frac{d\hat{Q}(\omega,t)}{dt} = -K_2 \frac{\partial}{\partial\hat{Q}}(|E(\omega,t)|^2) \qquad \ldots\ldots (2)$$

where K_1 and K_2 are constants which may be chosen so as to obtain an acceptable compromise between speed and stability of the parameter adjustments. The simplest selection ($K_1 = K_2 = K$), after some algebra, yields

$$\frac{d\hat{H}(\omega,t)}{dt} = -2K.E(\omega,t).X^*(\omega,t) \qquad \ldots\ldots (3)$$

where, as usual, the asterisk denotes complex conjugation (changing the sign of the imaginary component). Equation (3) completes the definition of the parameter-tracking arrangement shown in Fig. 1.

This type of TFA is found to perform very well when it is used to test systems operating under open-loop conditions. This is to be expected, for it can be shown (4) that the average values of the components of $\hat{H}(\omega,t)$ converge in the steady-state to values given by

$$[\hat{H}(\omega,t)]_{ave} \rightarrow \frac{\hat{S}_{xy}(\omega)}{\hat{S}_{x}(\omega)} \qquad \ldots\ldots (4)$$

where $\hat{S}_{xy}(\omega)$ is an estimate of the cross power spectral density of x(t) and y(t), and $\hat{S}_{x}(\omega)$ is an estimate of the power spectral density of x(t).

Since Equation (4) closely resembles the well-known classical form of estimate used for transfer function estimation[2] it is not surprising that its statistical properties are also remarkably similar to those of the classical estimates, and depend in much the same way on factors which determine the accuracy, resolution, and data requirements of the estimation procedure.

Unfortunately, the TFA shown in Fig. 1 also shares with the classical estimation procedure its inability to estimate feedback system characteristics accurately in the presence of noise. The reason for this is simply that the feedback loop introduces correlation between the noise and the measured system input (see Fig. 2) and this often biases the estimates significantly. Clearly, this is a serious shortcoming in a TFA which is intended for in-situ measurements, although it should be mentioned that this type of bias does not arise significantly if sinusoidal inputs to the system are allowed. To avoid such sources of errors with non-sinusoidal test signals, the parameter-tracking system has been modified to the form shown in Fig. 3. Comparing this with Fig. 1, it will be noted that the new form of the TFA calls for three signals from the system under test, rather than two as previously. The third signal, r(t), is supposed to be an independently generated signal which is correlated with x(t) and y(t), but not with any noise components in the feedback loop. Such a signal would be the reference input r(t) shown in Fig. 2, or even some other signal which was closely correlated with it. Filtered components of x(t) and y(t) are correlated with correspondingly filtered components of r(t) at the inputs to the parameter-tracking loop, by means of an additional set of complex multipliers. The action of the tracking loop is now such that the complex gain estimate $\hat{H}(\omega,t)$ converges in the mean to the quantity

$$[\hat{H}(\omega,t)]_{ave} \rightarrow \frac{\hat{S}_{ry}(\omega)}{\hat{S}_{rx}(\omega)} \qquad \ldots\ldots (5)$$

where $\hat{S}_{ry}(\omega)$ and $\hat{S}_{rx}(\omega)$ are estimates of the cross-power spectral densities of r(t) with y(t) and x(t) respectively. It can be shown[2],[5] that this quantity, unlike the one defined by Equation (4), yields estimates of $\hat{H}(\omega)$ which are not biased by noise in the feedback loops of the system

3. Instrument Design Considerations

A prototype implementation of the system defined schematically in Fig. 3 has provided useful opportunities for the investigation of many problems of design and operation of the proposed TFA, and has also provided a useful means for the assessment of a number of proposals relating to the processing of analogue data by means of digital and hybrid circuit techniques.

3.1. Heterodyne Filters

Probably the most demanding requirements of the TFA arise in connection with the heterodyne filters by means of which three sets of signals are to be resolved and filtered into accurately matched narrow frequency bands. The three filters must have identical, stable, and accurately controllable centre frequencies and pass-band characteristics, and this is the main reason for the selection of a heterodyning technique to perform the narrow-band filtering operation- Basically, the heterodyning process translates signal components in the neighbourhood of the heterodyning frequency to a corresponding region around zero frequency, and the attenuation of components outside the pass band is controlled by the characteristics of low-pass filters which follow the sine/cosine resolvers.

The sine/cosine resolvers are of a special digitally controlled type which operates by switching precision resistances into the feedback paths of a set of operational amplifiers. The switching operations are controlled by the states of a counter in which the argument of the sine/cosine functions is stored as a binary coded number. By means of a special circuit organization [6] the device can resolve angular increments of $2\pi/N$ using only $\log_2 N$ stages involving precision resistances, operational amplifiers, and electronic switches, provided that N is made an integral power of 2.

Six low-pass filters were required, having a common cutoff frequency adjustable in the range 0.01 Hz to 10 Hz, accurately matched cutoff characteristics, and low drift. A hybrid digital technique was developed for this purpose, by means of which a set of closely matched fourth-order filters was realised with a common cutoff frequency which is locked to the digital clock frequency employed. The basic circuit employed is very similar to the conventional RC-type of low-pass network, except that a switched capacitance bridge is employed to simulate a resistance value proportional to the switching frequency. This type of circuit has many attractions in variable filter applications, and although additional guard filters were required to control possible aliasing effects, these were of very simple design since the frequency aliases were well separated from the filter pass band.

3.2. Reference Correlators

Two complex multiplier elements were required to correlate the filtered quantities $X(\omega,t)$ and $Y(\omega,t)$ with the filtered reference $R(\omega,t)$. However, it was found that accurate multipliers were not an essential requirement in this case, for it is known that the derivation of unbiased estimates from the relation of Equation (5) requires the cross spectra to be computed with respect to a common reference signal which needs only to be <u>strongly correlated</u> with the system reference $r(t)$ [8]. One such function is, in fact the signum function Sgn $(R(\omega,t)$, which takes values of ± 1 according to the Sign. of $R(\omega,t)$. Thus, the reference correlators were implemented as greatly simplified analogue switches operated according to the polarity of the filtered reference signal.

3.3. Parameter-tracking Loop Elements

The multiplier elements used to form the product $\hat{H}X$ were high-accuracy analogue devices, but those associated with the product X^*E in Fig. 3. were signum multipliers as in the case of the reference correlators. This simplification has been shown to have no adverse effect on the accuracy of the TFA, since the use of the signum product X* Sgn(e) in the feedback loop merely modifies the rate of convergence towards minimum error, rather than the actual location of the minimum. Other linear elements in the parameter loops (such as summers, integrators, etc) were constructed of simple linear electronic components.

3.4. Digital Sequencing and Clock Controls

The prototype TFA was arranged to operate in a swept-frequency mode, so that transfer function loci could be automatically plotted over any parts of the selectable ranges 0.01 to 20, 0.1 to 200, or 1.0 to 2000 Hz. To accomplish this, a 3.2 decade (4-2-2-1) counter was arranged to increment up or down in steps of 0, 1, 2, 5, 10, 20, 50, or 100 digits. Each binary weight of this counter was arranged to control resistance values in a Wien-bridge oscillator so that the generated frequency was proportional to the counter contents. The oscillator output was appropriately shaped to drive the TTL stages of the sine/cosine resolvers.

Further logic circuits were employed for such purposes as measurement interval selection, start/stop/hold controls, and sample-and-hold of the parameter estimates $\hat{P}(\omega)$ and $\hat{Q}(\omega)$

4. Results and Conclusions

The TFA described here has been tested on real and simulated systems of various kinds with moderately satisfactory results. Probably the most severe criticisms applicable at this time are :-

(i) As currently implemented, the instrument is difficult to set up for optimum dynamic range. This could probably be improved greatly if automatic ranging techniques were introduced so as to maintain optimum signal levels at the filters and multiplier outputs.

(ii) The speed of measurement of the instrument when using sinusoidal test signals compares favourably with that of a conventional TFA. With random test signals, however, measurement times are very long because of the frequency-sweeping technique which is involved. In this respect, the TFA compares unfavourably with numerical Fourier transform machines, and some means should be found either to speed up the parameter tracking processes (possibly by frequency translation in the data) or to permit simultaneous operation (or something approaching this) at a suitable number of frequencies.

Against these limitations may be set the capabilities of high resolution and dynamic range in excess of those possessed by numerically based systems, and a cost/performance potential which compares favourably with systems requiring complex analogue/digital interfaces.

5. References

(1) Elsden, C.S., and A.J. Ley, 'A Digital Transfer Function Analyzer Based on Pulse-rate Techniques'. Automatica, 5, No.1, 1969 (pp. 51 - 60).

(2) Hughes, M.T.G. and P.E. Wellstead, 'Accuracy of Transfer Function Estimates Derived from Numerical Fourier Transforms'. This conference.

(3) Hughes, M.T.G. and R. May, 'Estimation of Frequency Response from Random Data'. Proc. of IFAC Symp. Prague, 1967.

(4) May, R., 'Spectral Analysis of Engineering Systems'. Ph.D. thesis, University of Warwick, 1968.

(5) Robertson, J.R., 'A New Transfer Function Analyzer', Ph.D. thesis, University of Warwick, 1971.

(6) Hughes, M.T.G. and J.R. Robertson, 'A Digitally Controlled Sine/Cosine Resolver'. To be published.

(7) Robertson, J.R. 'A Capacitor-ratio Commutated Filter'. To be published.

(8) Akaike, H. 'Some Problems in the Application of the Cross-spectral Method. Published in 'Spectral Analysis of Time Series', Ed. Harris. Wiley 1967 (pp. 81-107).

6. Acknowledgement

The authors wish to acknowledge support received from the Science Research Council and from the Ford Motor Company for the research reported here.

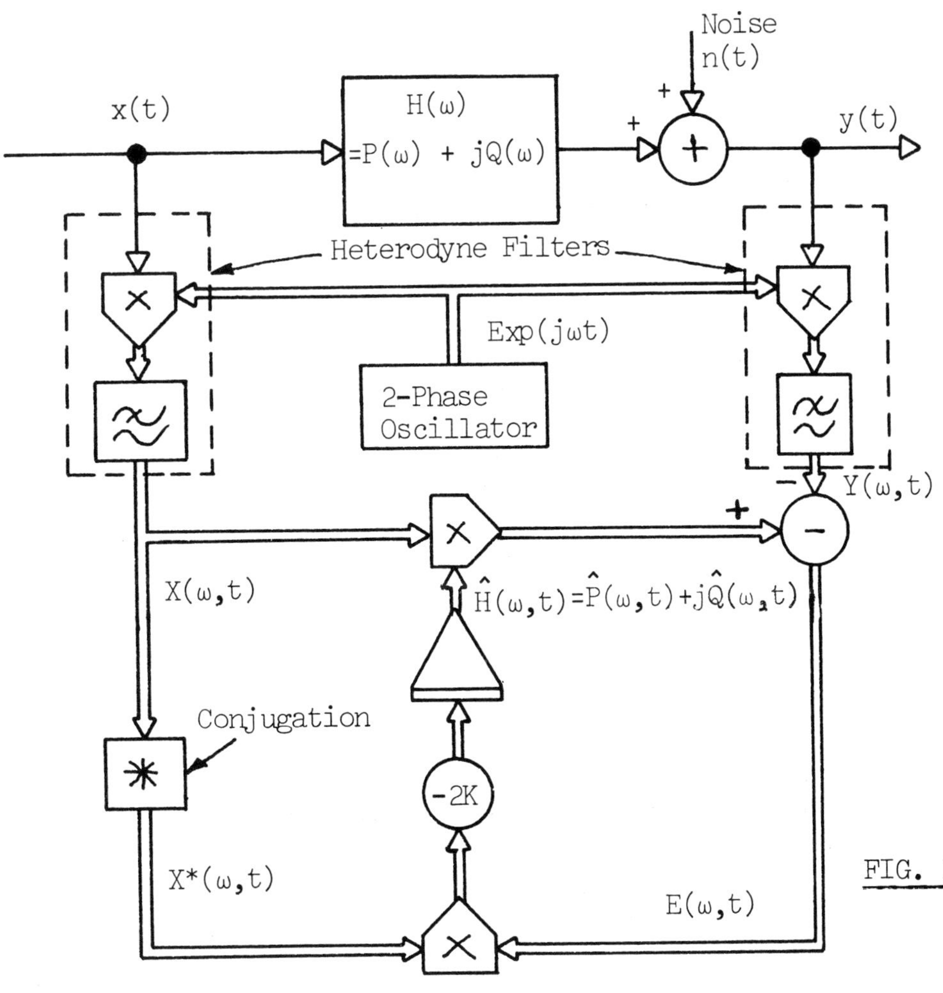

FIG. 1

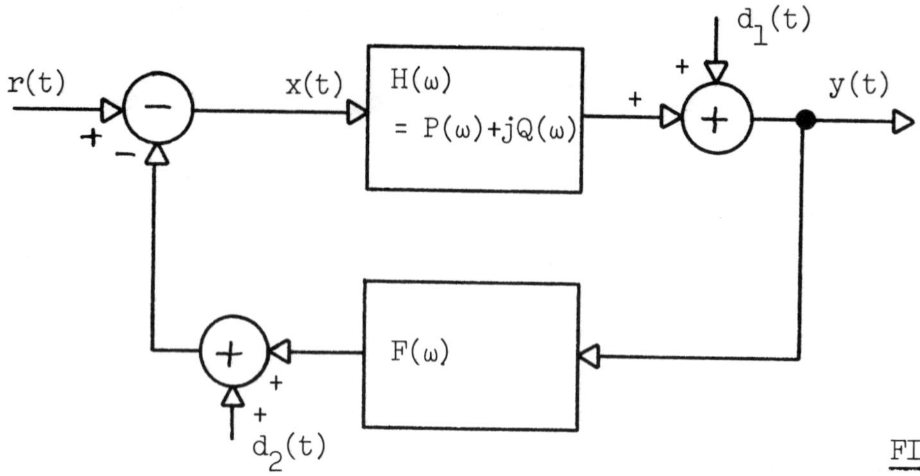

FIG. 2

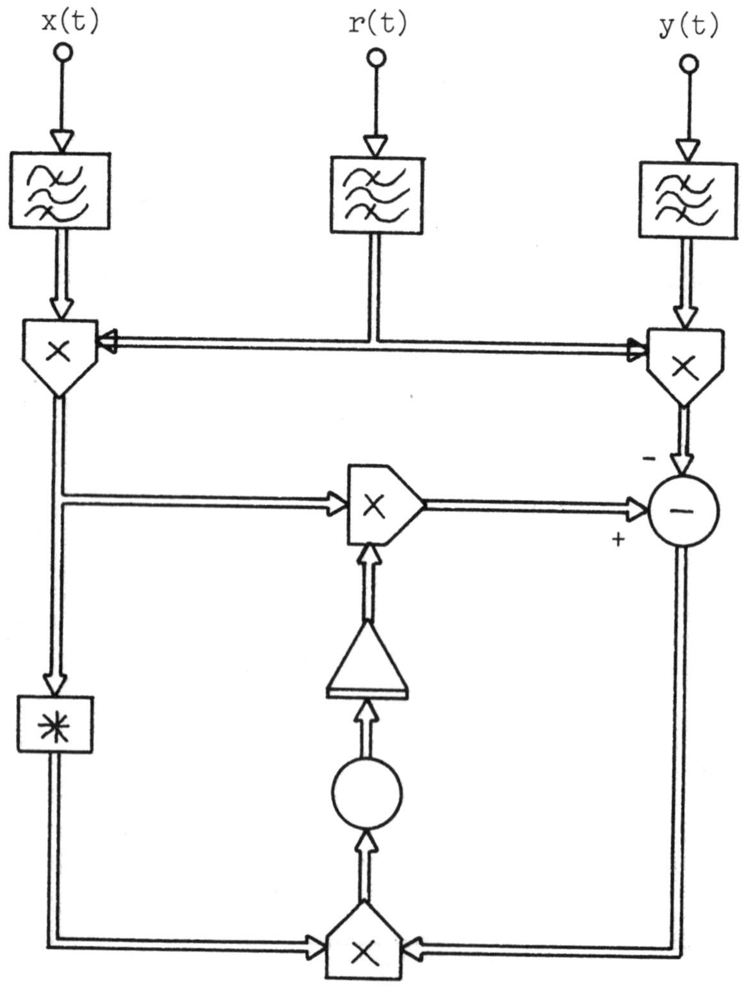

FIG. 3

A WALSH ANALYZER FOR PHASE AND AMPLITUDE MEASUREMENTS ON NETWORKS AND SYSTEMS.

W.M.Walmsley* and W.A.Evans*

1. INTRODUCTION :

Walsh functions form a complete set of orthogonal functions which, according to Fowle(1) were first recognised by Barrett around 1900, although Walsh (2) first introduced them into mathematics in 1923. Communications engineers were first to recognise the potential of using Walsh functions for digital data transmission (3), and to date the vast majority of practical applications are still in the areas of data transmission and multiplexing. Their use in instrumentation for frequency response analysis stems from the potential they offer in extending the upper frequency response of these instruments whilst retaining the low frequency capability of existing all-digital instruments. Presently available instruments operate in the Fourier domain and contain a swept sinusoidal signal source and a means for correlation and averaging, i.e. the machine responds as a narrow band filter rejecting noise. Frequency coverage extends over many decades ($10^{-4} \leqslant f \leqslant 10^{4}$ hertz). Results are normally available in cartesian (A +jB), polar (Z, angle θ) and log-polar (log Z, angle θ) co-ordinates. Methods vary as to how the correlation is performed, though the trend toward all-digital or hybrid instruments predominates (4,5).

In general, Walsh functions are defined over the unit interval θ = (0,1), i.e. any point on the function that occurs at a time t can be expressed as a fraction of the period T of the function, so that the value of $\theta (= t/_T)$ lies between 0 and 1. The functions over this interval may alternate between values of +1 and -1. Fig.2. illustrates the first sixteen Walsh functions. A detailed study of the schematic shows Wal(0,θ) to Wal(4,θ) can be identified with d.c., sin2πθ, cos2πθ, sin 4πθ and cos4πθ in terms of regularity of appearance and number of zero crossings over the interval θ. It will be appreciated that it is difficult to conceive an equivalent "frequency" for Walsh functions, and for this reason the term "sequency" has been adopted where sequency (zero crossings per sec) is equivalent to half the average number of zero crossings per sec. The concept of sequency (in z.p.s.) reduces to the same numerical value as that of frequency (in c.p.s.) when functions such as sinusoids are considered, and may be thought of as a more generalized concept of frequency applicable to functions that do not have an apparently well behaved appearance over any particular interval.

2. INSTRUMENT PHILOSOPHY :

The principle behind any instrument performing amplitude and phase measurements in the frequency domain is based on the orthogonal properties of the cisoidal group of waves. For a linear network or system excited by a sinusoidal wave the normalized output f(θ) may be written as :

$$f(\theta) = a_o + \sum_{n=1}^{\infty} a_n \cos(2\pi n\theta) + \sum_{n=1}^{\infty} \sin(2\pi n\theta) \qquad \ldots.(1)$$

where

$$a_n \doteq 2\int_0^1 f(\theta).\cos(2\pi n\theta).\,d\theta, \quad b_n = 2\int_0^1 f(\theta).\sin(2\pi n\theta).d\theta. \qquad \ldots.(2)$$

*The authors are with the Department of Electrical and Electronic Engineering, University College of Swansea, SA2 8PP.

If the exciting input is $\sin 2\pi\theta$ then the output will contain a term $R \sin(2\pi\theta + \phi)$ where :

$$R = (a_1^2 + b_1^2)^{1/2} \quad \text{and} \quad \phi = \arctan b_1/a_1 \qquad \ldots.(3)$$

The instrument is designed to perform operations indicated in equations (2) and (3). It may be shown (4) that the equivalent filter response is given by :

$$\left| \mathrm{Re}\left[H(j/\theta)\right] \right| = \frac{2\sin N\pi/\theta}{\pi N(1-1/\theta^2)}, \qquad \left| \mathrm{Im}\left[H(j/\theta)\right] \right| = 1/\theta \left| \mathrm{Re}\left[H(j/\theta)\right] \right| \qquad \ldots.(4)$$

The envelope of these curves has a general shape common to all analyzers that perform the integral expressions contained in equation (2). Fig.3. is an example of a plot of the in-phase component given by the first integral as a function of the continuous variable $\omega/\omega_0 = 1/\theta$. An additional parameter N has been introduced indicating that the integral has been taken over N instead of 1 cycle as in equation (2). The effect of increasing N narrows the equivalent filter bandwidth giving greater noise rejection, as shown in the schematic.

The above analysis has been confined to expanding $f(\theta)$ as a series of sines and cosines. It equally follows that the same process could be repeated with any set of functions having orthogonal properties - negative and positive exponentials, Bessel functions and Lagrange polynomials are familiar examples. In a strictly practical situation however, we are confined to using functions that can easily be generated with hardware, if an alternative approach is to be adopted. A closer look at the physical problem indicates that the cisoidal group of functions contains an additional unique property in that if they are used to excite a linear system they remain unchanged except in amplitude and phase. It follows that the choice of excitation already appears to be the best one, and the use of alternative orthogonal functions must rest with the ease with which the computational process can be carried out and the conversion from the chosen orthogonal set to the cisoidal set. The procedure for the Walsh set is graphically shown in Fig.1. In the Fourier domain both time and frequency plots of the exciting sinusoid are shown. Alternatively, the sinusoid may be first expanded as a series of Walsh functions according to the general expression :

$$f(\theta) = A_0 + \sum_{m=1}^{\infty} A_m \mathrm{cal}(m,\theta) + \sum_{m=1}^{\infty} B_m \mathrm{sal}(m,\theta) \qquad \ldots.(5)$$

to give an odd series of Walsh functions. Each Walsh function in turn may be expanded as a Fourier series according to the general expression of equation (1), and the results at each frequency harmonic combined using the principle of superposition. Siemens (6) has shown that there exists transformation matrices which map from the Walsh domain to the Fourier domain yielding relationships between the respective coefficients.

In the present context it is more appropriate to express $f(\theta)$ by a single Walsh or exponential series respectively.

$$\text{i.e.}\quad f(\theta) = \sum_{m=0}^{2M-1} W_m . \mathrm{Wal}(m,\theta) = \sum_{n=0}^{N-1} F_n . e^{j2\pi n\theta} . \qquad \ldots(7)$$

The Wal term may also be expanded as an exponential series

$$\mathrm{Wal}(m,\theta) = \sum_{n=0}^{N-1} T_{n,m}\, e^{j2\pi n\theta} . \qquad \ldots(8)$$

Substituting (8) in (7) and equating coefficients of the respective exponential terms :

$$[F] = [T] \cdot [W] \qquad \ldots(9)$$

$[F]$ and $[T]$ are complex matrices from which a_n and b_n may be derived :

$$a_n = \mathrm{Re}\{[T]\} \cdot [W] , \qquad b_n = I_m\{[T]\} \cdot [W] \ldots(10)$$

Provided $2M = N = 2^x$ (where x is an integer > 0), then the first N Fourier coefficients may be obtained without truncation error.

3.0. DESIGN RATIONALE :

Before entering into a more detailed account of the instrument, certain fundamental questions require answering. It is necessary to establish what word length the instrument will use, since the wrong choice would yield a result limiting the upper frequency response with attendant increase in hardware costs. Series truncation errors in the expansion of the exciting sinusoid into a series of Walsh functions, together with computation carried out with finite word lengths, are two sources of error. The errors are distributed from both sources, sometimes enhancing each other and sometimes cancelling. A simulation study of the overall machine has been carried out for this purpose to yield information on the anticipated performance of the instrument. The filter responses in Fig.5. have been plotted for single cycle integration for various word and transform lengths. The plots establish that an eight bit word length constitutes an admirable compromise between speed, accuracy and economy of hardware. Together with a sixteen point transform, the eight bit word length has formed the basic design specification of the instrument. Fig.4. shows the upper bounds of anticipated errors expressed as standard deviations for various word lengths with transform length as the running parameter.

The basic requirements of any Walsh transform logic is simply a capability to add, subtract and perform the logical shift right. The design has been proved under simulated conditions and the transform can be completed using standard TTL in less than 16 microseconds (with sufficient time to time share the processor for other operations).

The Walsh Fourier conversion is performed in two stages. The Walsh coefficients ω_0 to $_{15}$ are treated as two distinct sets of numbers :

$$(W_0, W_2, W_4, \ldots W_{14}) \qquad (W_1, W_3, W_5, \ldots W_{15})$$

The even coefficients are used to realise the even (real) Fourier coefficients (a) and the odd coefficients may be processed to yield the odd (imaginary) Fourier coefficients (jb). The ROM storage necessary for the conversion constants amounts to only thirteen eight -bit words (13 'bytes'). Logic for the noise rejecting averaging routine is also required to reduce the contribution of uncorrelated signals. The display algorithms are carried out by adapting a desk calculator chip for the purpose.

4.0. CONCLUSIONS :

The preliminary work carried out so far demonstrates that the use of orthogonal functions other than the cisoidal group is not only feasible but attractive. A substantial, if not dramatic, reduction in hardware costs is indicated, combined with extended high frequency performance (~60 kHz) as well as serial/parallel operation. The instrument has an excitation section in common with existing all-digital instruments. To summarise, the instrument retains the advantages of existing machines down to the lowest frequencies if necessary, whilst extending the upper frequency and including information up to the seventh harmonic in a parallel mode of operation. This is particularly useful for certain types of non-linear systems where linear methods are used to obtain information about the system.

5.0. REFERENCES :

1. FOWLE, F.F. "The Transposition of Conductors" Trans. AIEE (1905) 23, pp.659-687.

2. WALSH, J.L. "A Closed Set of Orthogonal Functions". Amer. Jour. Math. (1923) 45 pp.5-24.

3. HARMUTH, H.F. "Application of Walsh Functions in Communications" IEEE Spectrum (1969). Nov., pp.82-91.

4. ELSDEN, C.S. and LEY, A.S. "A Digital Transfer Function Analyzer based on Pulse Rate Techniques". Automatica (1969) 5, 1, pp.51-60.

5. PAULL, C.J. and EVANS, W.A. "The Application of a Commutated Filter to the Design of a Frequency Response Analyzer". The Radio & Electronic Engineer (1973), 43, 6, pp.369-378.

6. SIEMENS, K.H. "Digital Walsh-Fourier Analyzer for Periodic Waveforms". M.Eng.Thesis 1969, McMaster University, Ontario, Canada.

6. ACKNOWLEDGEMENT :

The authors wish to acknowledge the financial and technical assistance of The Wayne Kerr Company Limited in support of this work.

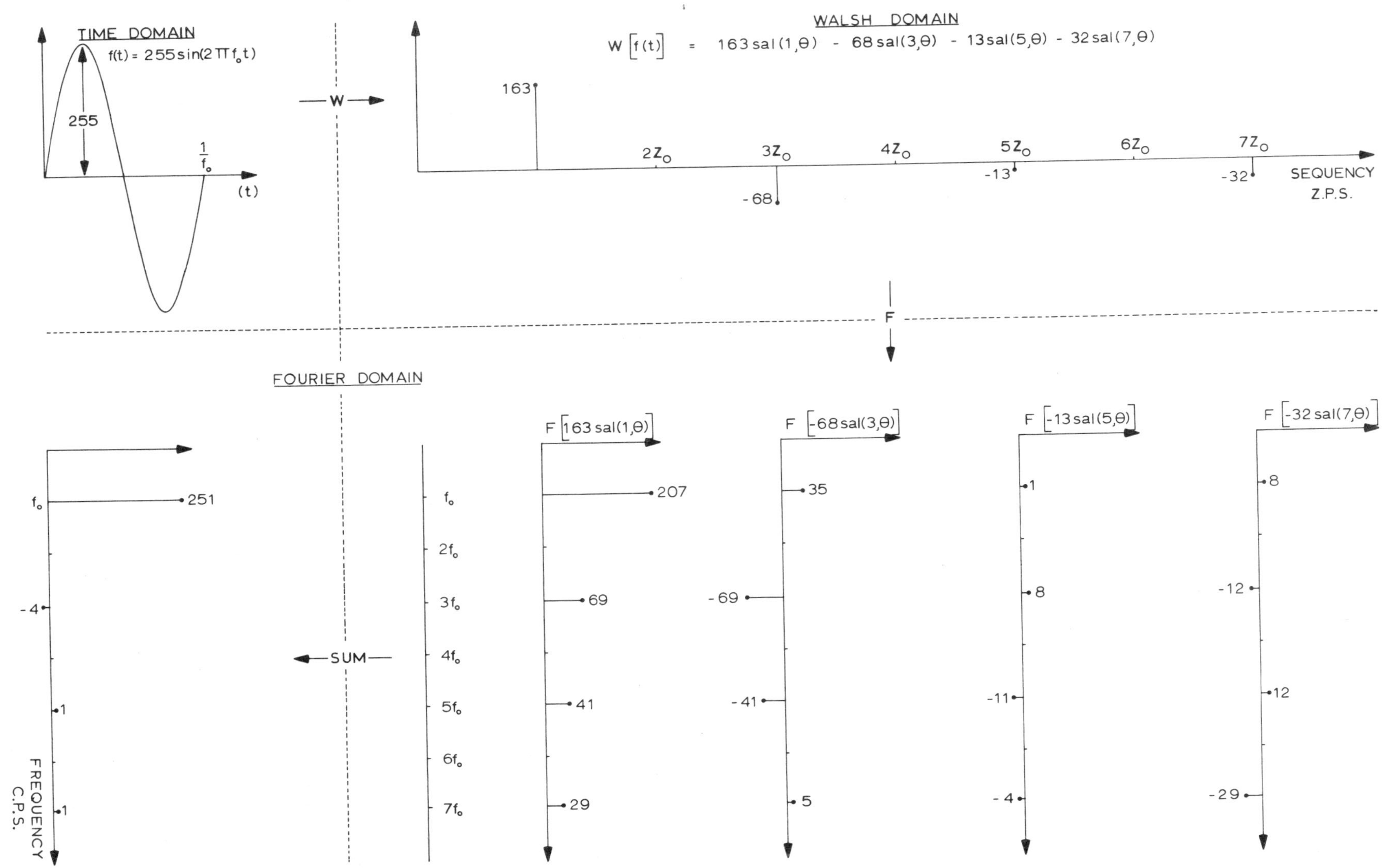

FIG 1 PRINCIPLE OF SUPERPOSITION APPLIED TO WALSH-FOURIER TRANSFORMS

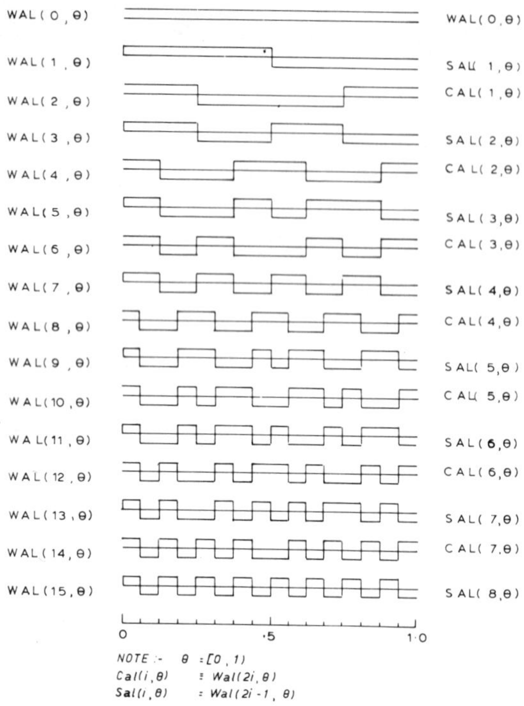

FIG 2 WALSH FUNCTIONS

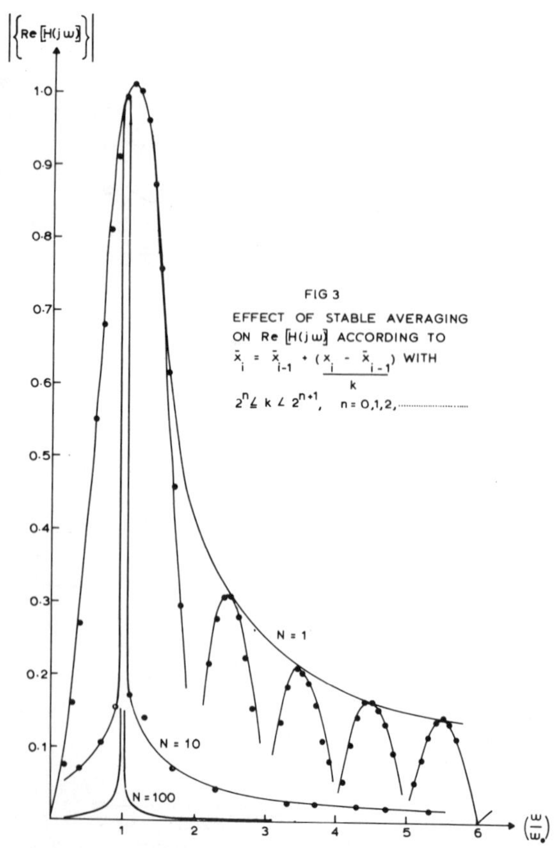

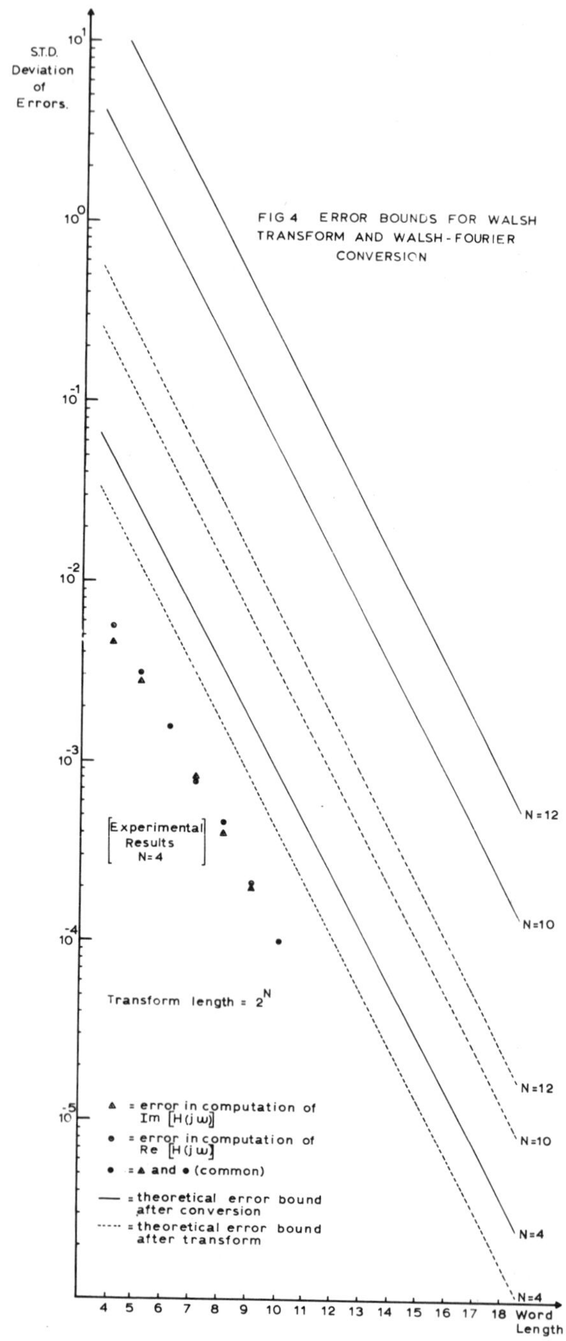

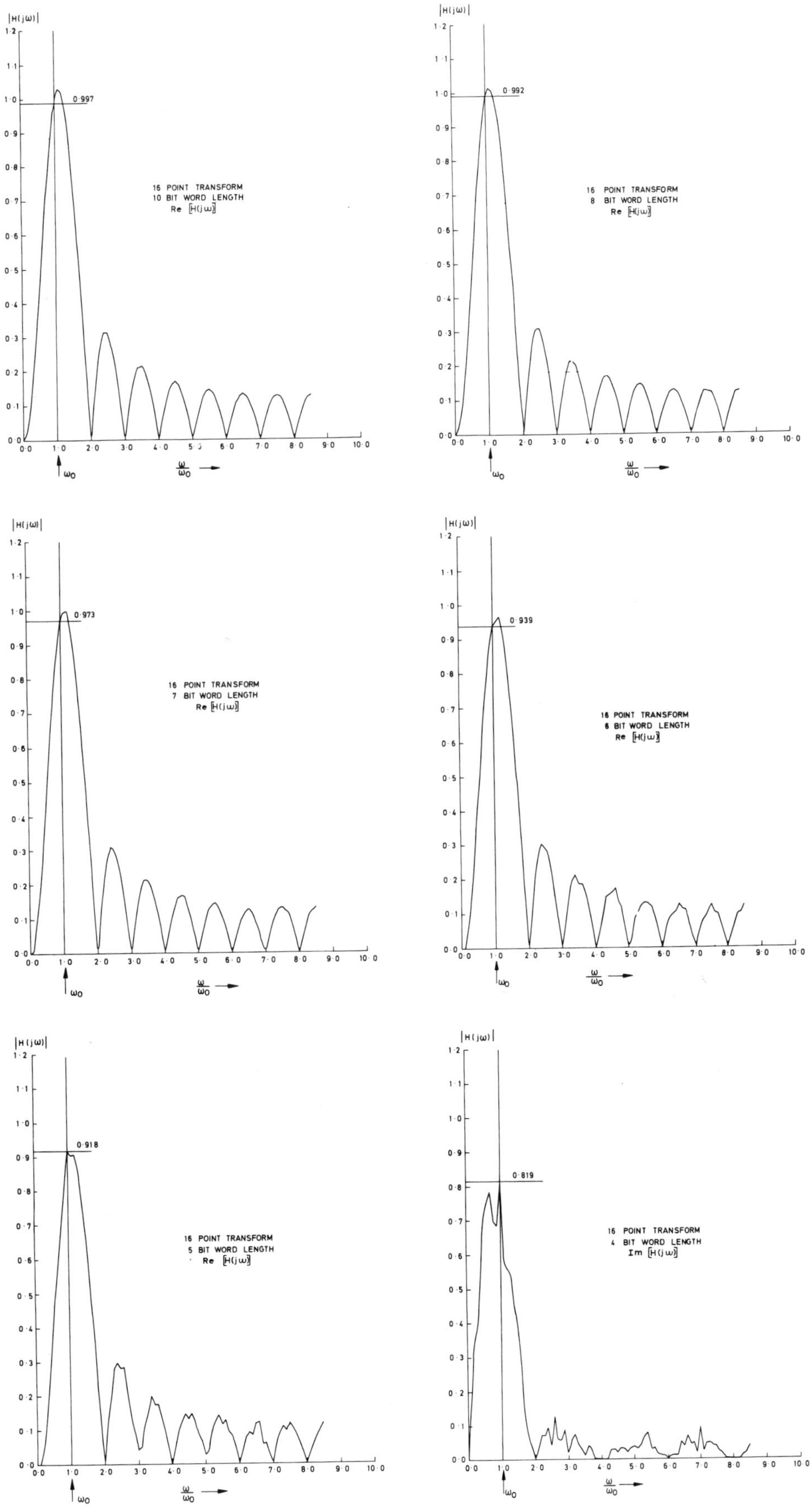

FIG. 5 EQUIVALENT FILTER RESPONSES WITH SINGLE CYCLE INTEGRATION AND WALSH DOMAIN PROCESSING.